Lecture Notes in Computer Science   12516

More information about this subseries at http://www.springer.com/series/7407

Tristan Cazenave · Jaap van den Herik ·
Abdallah Saffidine · I-Chen Wu (Eds.)

Advances in Computer Games

16th International Conference, ACG 2019
Macao, China, August 11–13, 2019
Revised Selected Papers

 Springer

Editors
Tristan Cazenave ⓘ
Université Paris-Dauphine
Paris, France

Jaap van den Herik ⓘ
Leiden University
The Hague, The Netherlands

Abdallah Saffidine ⓘ
The University of New South Wales
Sydney, NSW, Australia

I-Chen Wu ⓘ
National Chiao-Tung University
Hsin-Chu, Taiwan

ISSN 0302-9743   ISSN 1611-3349   (electronic)
Lecture Notes in Computer Science
ISBN 978-3-030-65882-3   ISBN 978-3-030-65883-0   (eBook)
https://doi.org/10.1007/978-3-030-65883-0

LNCS Sublibrary: SL1 – Theoretical Computer Science and General Issues

This Springer imprint is published by the registered company Springer Nature Switzerland AG
The registered company address is: Gewerbestrasse 11, 6330 Cham, Switzerland

Preface

This book contains the papers of the 16th Advances in Computer Games Conference (ACG 2019) held in Macau, China. The conference took place during August 11–13, 2019, in conjunction with the IJCAI conference, the Computer Olympiad, and the World Computer-Chess Championship.

The ACG conference series is a major international forum for researchers and developers interested in all aspects of artificial intelligence and computer game playing. Earlier conferences took place in London, UK (1975), Edinburgh, UK (1978), London, UK (1981, 1984), Noordwijkerhout, The Netherlands (1987), London, UK (1990), Maastricht, The Netherlands (1993, 1996), Paderborn, Germany (1999), Graz, Austria (2003), Taipei, Taiwan (2005), Pamplona, Spain (2009), Tilburg, The Netherlands (2011), and Leiden, The Netherlands (2015, 2017).

In this conference 19 papers were submitted. Each paper was sent to three reviewers. The Program Committee accepted 12 papers for presentation at the conference and publication in these proceedings. As per usual, we informed the authors that they submitted their contribution to a post-conference editing process. The two-step process is meant (a) to give authors the opportunity to include the results of the fruitful discussion after the lecture in their paper, and (b) to maintain the high-quality threshold of the ACG series.

Moreover three invited talks were given by Jonathan Schaeffer, Nathan Sturtevant and Cameron Browne, Eric Piette, and Matthew Stephenson.

We now give a brief introduction to the papers included in this book regrouped by topics.

Cooperation

The first paper is "Advice are Useful for Game AI: Experiments with Alpha-Beta Search Players in Shogi" by Shogo Takeuchi. It presents methods to strengthen a game AI using advice from other game AIs during game play. Advice are moves selected by an adviser and propose a mechanism that makes a player search again when the player's move is different from advice. Experiment are made for the game of Shogi.

The second paper by Eisuke Sato and Hirotaka Osawa is "Reducing Partner's Cognitive Load by Estimating the Level of Understanding in the Cooperative Game Hanabi." Hanabi is a cooperative game for ordering cards through information exchange. Cooperation is achieved in terms of not only increased scores, but also reduced cognitive load for the players. The thinking time is used as an indicator of cognitive load, and the results showed that it is inversely proportional to the confidence of choice. When the agent uses the thinking time of the player the mean thinking time of the human player is shortened. It suggests that it could reduce the cognitive load of the players without influencing performance.

The third paper by Gregory Schmidt and Philip Shoptaugh is "Making a Better Game: The History of Cluster." The authors present a case study of the initial inspiration and design process that led to successfully optimized versions of the game Cluster.

Single Player Games

The fourth paper by Taishi Oikawa, Chu-Hsuan Hsueh, and Kokolo Ikeda is "Enhancing Human Players' T-Spin Technique in Tetris with Procedural Problem Generation." They are interested in programs that can entertain or teach human players. They automatically generate puzzles so that human players improve at playing the game of Tetris. A technique hard to learn for beginners is T-spin. Automatically generated two-step T-spin problems are given to human players to solve and they enable to improve their skills at Tetris.

The fifth paper by Kiminori Matsuzaki is "A Further Investigation of Neural Network Players for Game 2048." Game 2048 is a stochastic single-player game. Strong 2048 computer players use N-tuple networks trained by reinforcement learning. The paper investigates neural-network players for Game 2048 and improve their layers and their inputs and outputs. The best neural-network player achieved an average score of 215 803 without search techniques, which is comparable to N-tuple-network players.

Mathematical Approaches

The sixth paper by Michael Hartisch and Ulf Lorenz is "A Novel Application for Game Tree Search - Exploiting Pruning Mechanisms for Quantified Integer Programs." They investigate pruning in search trees of so-called quantified integer (linear) programs (QIPs). QIPs consist of a set of linear inequalities and a minimax objective function, where some variables are existentially and others are universally quantified. They develop and theoretically substantiate tree pruning techniques based upon algebraic properties. The implementation of their findings can massively speed up the search process.

The seventh paper by Nicolas Fabiano and Ryan Hayward is "New Hex Patterns for Fill and Prune." A fill pattern in the game of Hex is a subposition with one or more empty cells that can be filled without changing the position's minimax value. Some cells can be pruned and ignored when searching for a winning move. They introduce two new kinds of Hex fill – mutual and near-dead – and some resulting fill patterns. They show four new permanently-inferior fill patterns and present three new prune results, based on strong-reversing, reversing, and game-history respectively.

The eighth paper by Jos Uiterwijk is "Solving Cram Using Combinatorial Game Theory." He investigates the board game Cram, which is an impartial combinatorial game, using an $\alpha\beta$ solver. He uses knowledge obtained from Combinatorial Game Theory (CGT) for his solver. Using endgame databases pre-filled with CGT values (nimbers) for all positions fitting on boards with at most 30 squares and also using two efficient move-ordering heuristics gives a large improvement of solving power. He also defines five more heuristics based on CGT that further reduce the sizes of the solution trees considerably. He was able to solve all odd by odd Cram boards for which results were available from the literature (even by even and odd by even boards are trivially solved). He proves new results for the 3×21 board, a first-player win, and the 5×11 board, a second-player win.

Nonogram: General and Specific Approaches

The ninth paper by Aline Hufschmitt, Jean-Noel Vittaut, and Nicolas Jouandeau is "Exploiting Game Decompositions in Monte Carlo Tree Search." They propose the Multiple Tree MCTS (MT-MCTS) approach to simultaneously build multiple MCTS trees corresponding to different sub-games. They apply it to single player games from General Game Playing. Complex compound games are solved from 2 times faster (Incredible) up to 25 times faster (Nonogram).

The tenth paper by Yan-Rong Guo, Wei-Chiao Huang, Jia-Jun Yeh, Hsi-Ya Chang, Lung-Pin Chen, and Kuo-Chan, Huang is "On Efficiency of Fully Probing Mechanisms in Nonogram Solving Algorithm." Fully probing plays is important for Nonogram. The authors address several critical factors influencing fully probing efficiency: re-probing policy, probing sequence, and computational overhead. Taking into account these factors improves the speed of solving Nonogram puzzles significantly.

Deep Learning

The eleventh paper by Hsiao-Chung Hsieh, Ti-Rong Wu, Ting Han Wei, and I-Chen Wu is "Net2Net Extension for the AlphaGo Zero Algorithm." The number of residual blocks of a neural network that learns to play the game of Go following the AlphaGo Zero approach is important for the strength of the program but also takes more time for self-play. The authors propose a method to deepen the residual network without reducing performance. The deepening process is performed by inserting new layers into the original network. They present three insertion schemes. For 9×9 Go, they obtain a 61.69% win rate against the unextended player while greatly saving the time for self-play.

The twelfth paper by Tomihiro Kimura and Kokolo Ikeda is "Designing Policy Network with Deep Learning in Turn-Based Strategy Games." They apply deep learning to turn-based strategy games. A recurrent policy network is developed learning from game records. The game data are generated using Monte Carlo Tree Search. The resulting policy network outperforms MCTS.

Invited Papers

The first invited paper by Nathan Sturtevant is "Steps towards Strongly Solving 7x7 Chinese Checkers." Chinese Checkers is a game for 2–6 players that has been used as a testbed for game AI in the past. He provides an overview of what is required to strongly solve versions of the game, including a complete set of rules needed to solve the game. We provide results on smaller boards with result showing that these games are all a first-player win.

The second invited paper by Cameron Browne, Matthew Stephenson, Éric Piette, and Dennis J.N.J. Soemers is "The Ludii General Game System: Interactive Demonstration." Ludii is a new general game system, currently under development, which aims to support a wider range of games than existing systems and approaches. It is being developed primarily for the task of game design, but offers a number of other potential benefits for game and AI researchers, professionals, and hobbyists. The paper describes the approach behind Ludii, how it works, how it is used, and what it can potentially do.

Acknowledgment

This book would not have been produced without the help of many persons. In particular, we would like to mention the authors and reviewers for their help. Moreover, the organizers of IJCAI, the Computer Olympiad, and the World Computer-Chess Championship have contributed substantially by bringing the researchers together.

September 2020

Tristan Cazenave
Jaap van den Herik
Abdallah Saffidine
I-Chen Wu

Organization

Executive Committee

Tristan Cazenave
Jaap van den Herik
Abdallah Saffidine
I-Chen Wu

Organizing Committee

Grace Krabbenbos
Johanna Hellemons
Hiroyuki Iida
Jan Krabbenbos
Tristan Cazenave
Jaap van den Herik
Abdallah Saffidine
I-Chen Wu
David Levy

Program Committee

Jean-Marc Alliot	Institut de Recherche en Informatique de Toulouse, France
Hendrik Baier	Digital Creativity Labs, University of York, UK
François Bonnet	Tokyo Institute of Technology, Japan
Bruno Bouzy	Paris Descartes University, France
Cameron Browne	Queensland University of Technology, Australia
Marc Cavazza	University of Greenwich, UK
Tristan Cazenave	LAMSADE, Université Paris Dauphine, PSL, CNRS, France
Lung-Pin Chen	Tunghai University, Taiwan
Tzung-Shi Chen	National University of Tainan, Taiwan
Albert Chen	Yonsei University, South Korea
Siang Yew	University of Nottingham, Malaysia Campus, Malaysia
Hsin-Hung Chou	Chang Jung Christian University, Taiwan
Vincent Corruble	LIP6, Sorbonne University Pierre and Marie Curie Campus, France
Johannes Fürnkranz	TU Darmstadt, Germany
Chao Gao	University of Alberta, Canada
Reijer Grimbergen	Tokyo University of Technology, Japan

Michael Hartisch	University of Siegen, Germany
Ryan Hayward	University of Alberta, Canada
Jaap van den Herik	Leiden University, The Netherlands
Rania Hodhod	Columbus State University, USA
Christoffer Holmgård	Northeastern University, USA
Tsan-Sheng Hsu	Academia Sinica, Taiwan
Chun-Chin Hsu	Chang Jung Christian University, Taiwan
Chu-Hsuan Hsueh	Japan Advanced Institute of Science and Technology, Japan
Kuo-Chan Huang	National Taichung University of Education, Taiwan
Hiroyuki Iida	Japan Advanced Institute of Science and Technology, Japan
Takeshi Ito	The University of Electro-Communications, Japan
Eric Jacopin	CREC Saint-Cyr, France
Nicolas Jouandeau	University Paris 8, France
Tomoyuki Kaneko	The University of Tokyo, Japan
Hung-Yu Kao	National Cheng Kung University, Taiwan
Akihiro Kishimoto	IBM Research, Ireland
Walter Kosters	LIACS, Leiden University, The Netherlands
Sylvain Lagrue	Universitè de Technologie de Compiégne (UTC), France
Xuejun Li	Anhui University, China
Shun-Shii Lin	National Taiwan Normal University, Taiwan
Jialin Liu	Southern University of Science and Technology, China
Richard Lorentz	California State University, USA
Kiminori Matsuzaki	Kochi University of Technology, Japan
Helmut Mayer	University of Salzburg, Austria
Makoto Miwa	Toyota Technological Institute, Japan
Martin Mueller	University of Alberta, Canada
Todd Neller	Gettysburg College, USA
Mark J. Nelson	American University, USA
Diego Perez Liebana	Queen Mary University of London, UK
Mike Preuss	Leiden University, The Netherlands
Aleksander Sadikov	University of Ljubljana, Slovenia
Abdallah Saffidine	University of New South Wales, Australia
Jahn-Takeshi Saito	Lesson Nine GmbH, Germany
Spyridon Samothrakis	University of Essex, UK
Maarten Schadd	TNO, The Netherlands
Jonathan Schaeffer	University of Alberta, Canada
Moshe Sipper	Ben-Gurion University of the Negev, Israel
Nathan Sturtevant	University of Alberta, Canada
Ruck Thawonmas	Ritsumeikan University, Japan
José Valente de Oliveira	Universidade do Algarve, Portugal
Jonathan Vis	Leiden University Medical Center, The Netherlands
Tinghan Wei	National Chiao Tung University, Taiwan
Mark H. M. Winands	Maastricht University, The Netherlands

Contents

Advice is Useful for Game AI: Experiments with Alpha-Beta Search Players in Shogi

Shogo Takeuchi[✉]

Kochi University of Technology, Kami, Japan
takeuchi.shogo@kochi-tech.ac.jp

Abstract. In this paper, we present methods to strengthen a game AI using advice from other game AIs during game play. People can improve their strength with advice, such as finding better moves or avoiding mistakes. Therefore, to improve the performance of game AIs, we focused on advice. The issues to be considered are the definition of advice and mechanism with advice for move selection. In this paper, we propose that "advice" are moves selected by an adviser and propose a mechanism that makes a player search again when the player's move is different from advice. We performed tournaments among the proposed systems and other methods against a single engine to compare the strength in shogi. We showed the effectiveness of the proposed method from the experimental results and demonstrated that game AIs can improve their strength with advice. In addition, we found that the advice from a weaker game AI is still useful for game AI.

Keywords: Search · Advice · Game

1 Introduction

AlphaGo (Zero) has attracted much attention in the game of go because of its effective learning algorithm and strong results in the man-machine matches [11]. Games have become a testbed of artificial intelligence, particularly, of intelligent engines that can win against human champions. Strong engines require a search method and a positional evaluation scheme. Alpha–beta search assisted by an evaluation function [9] has been the most successful approach in chess and its Japanese variant, shogi. Recently, a new approach by AlphaGo (Zero), Monte-Carlo Tree Search with a deep neural network, has been proven successful not only in the game of go but also in chess and shogi [10].

Advice is forbidden in game tournaments among human players. Humans can improve their strength by finding better moves or avoiding mistakes with advice. This applies to advice from human to human; what about other cases? We are interested in enhancing the strength of game AIs. Thus, we focus on

This work was supported by JSPS Grant-in-Aid for Young Scientists (A) No. 17K12807.

T. Cazenave et al. (Eds.): ACG 2019, LNCS 12516, pp. 1–10, 2020.
https://doi.org/10.1007/978-3-030-65883-0_1

the advice from game AIs to game AIs. The research problem discussed in this paper is whether game AIs can improve their strength with advice. A definition of advice and its usage is required.

Advice from game AIs to human has been studied in advanced chess [3] and 3-Hirn system [2]. A human player uses a game AI to enhance his/her strength by avoiding making a mistake or assuring his/her thought in advanced chess. Game AIs suggest candidate moves to the human player and the human player selects one from them in the 3-Hirn system. Both results showed that human players improve their strength using game AIs. The candidate moves can be seen as advice from game AIs in 3-Hirn systems.

In a majority voting system, we use multiple game engines, and each engine votes to its search result (move); then, we select the best move by the number of votes [8]. If we view search results as advice, the system can be viewed as summarizing engines' advice.

The paper is structured as follows. In section two, we introduce related work about advice in games. We propose our method to improve the strength of game AI with advice from game AIs in section three. Section four presents the experimental evaluation of our methods. Finally, section five gives an overall view of this paper, leading to conclusions and future work.

2   Related Work

There are three types of studies on advice. One is to improve the learning process with advice. The other one is advice from game AIs to a human player. In addition, majority voting can be viewed as a method to utilize advice from and to game AIs.

Taylor et al. proposed a teacher-student framework for reinforcement learning and investigated the effect of the amount and moments of advice [15]. Their goal to make agents stronger by learning is different from ours. They improve the learning method with advice; however, we try to improve the AIs' decision with advice during gameplay. A teacher provides students with advice, that is, action in this framework.

2.1   Advice from Game AIs to Human

There are studies about advice from game AIs to humans, such as advanced chess [3,6] and 3-Hirn, Double-Fritz, and List-3-Hirn [1,2].

In advanced chess, a human player uses a game AI to enhance his/her strength by avoiding a mistake or assuring his/her thought.

In 3-Hirn, Double-Fritz, and List-3-Hirn, game AIs suggest candidate moves to a human player, and then a human player selects the best move from candidates. Differences between the three are as follows: Two different engines suggest their best moves to a human player in 3-Hirn, one engine suggests top-k moves to a human player in Double-Fritz, and two different engines suggest their top-k moves to a human player in List-3-Hirn. Even if the human player thinks of other moves as being better, he cannot select that move in this method.

It is not clear how human players utilize game AIs to improve their strength in advanced chess. Meanwhile, it is relatively clear in 3-Hirn and its variants. A human player receives candidate moves from game AIs and then selects one move from those. The candidate moves can be seen as advice in those methods.

2.2 Majority Voting

Obata et al. proposed a majority voting system in shogi [8]. In a majority voting system, engines do not share the information during the search, but the system summarizes their search results and selects the best after all engines have finished their search. Their experiment showed that the majority voting system is stronger than the single engines. If we view search results as advice, the system can be viewed as summarizing engines' advice.

Another voting approach in games is optimistic voting, which selects the move with the highest evaluation value [12]. Optimistic voting [12] is a simple idea that the move with the highest score is selected. This study was conducted for shogi with the homogeneous system, and alpha–beta search players.

Marcolino et al. showed that the diversity of the voting system plays an important role. They compared the homogeneous system with the heterogeneous system [7] and other parallelization methods. From their experiments, the heterogeneous system is stronger than the homogeneous system.

However, a majority voting with a heterogeneous system has a problem with weaker engines [7]. For example, if the weaker engines are the majority, then the votes from the stronger engines are ignored, and the system becomes weaker. Takeuchi proposed a weighted voting method in a heterogeneous system [13] to solve the problem.

3 Proposed Method

The research problem in this paper is whether game AIs can improve their strength with advice. Before beginning, a definition of advice and their usage are required.

First, we define what "advice" in the present paper is. Advice is a move (or an action) in the related work such as teacher-student framework [15] and 3-Hirn system [2]. Thus, we define "advice" as a candidate move in the context of present paper.

All engines, including a player, search and obtain their search results (best moves). If all moves are the same, that move is selected; otherwise, a player searches again and then selects one.

1. A player searches a position p and obtains the best move m (and its score v).
2. Each adviser e searches the same position p and obtains the best move $m_{e,a}$, "advice."

(a) If all advice $m_{e,a}$ are the same with the player's best move m, then, player selects the move m.

(b) Otherwise, the player searches the position p again with the same search time with the previous search and selects the best move.

Searching the same position can be seen as a search extension because game AIs use hash tables to store the previous search results.

There are options for the moves to search. Searching advice m_a only, advice m_a and its best move m, and all legal moves. If the search time is the same for each option, the shallower search goes deeper than the wider one. If the option selected is searching advice only, we need to compare the obtained score with the previous search result and select the move with a higher score.

3.1   Options of Searching Moves

We have three options below:

Advice a player searches advice only,
Both a player searches advice and its best move,
All a player searches all moves.

Now, we discuss when advice is effective and which option to choose. We consider two cases as follows:

case A m_a is better move (the adviser notices that), however the player scores the move lower.

case B m is worse move (the adviser selects the other), however the player scores the move the highest.

To select a better move, the player should notice that the m_a is better in case B; thus, the move m_a should be searched deeper. However, in case A, the player must notice that the selected move m is worse; therefore, the move m must be searched deeper.

Thus, option "Advice" improves case A and cannot improve case B, whereas other options can improve both cases. However, their search depth would be shallower than that of option "Advice". All options have a tradeoff between search width (covering both cases is better) and search depth (searching deeper is better). The frequency with which case A occurs will decrease if the adviser is weaker and increase if the adviser is stronger. Thus, the better options seem to depend on the strength of the adviser.

4   Experiments

We performed tournament experiments in the game of shogi to compare the performance of the proposed methods and optimistic voting. Shogi [5], which is similar to chess, is a popular game in Japan. The strongest shogi programs have recently defeated top human players [14] thanks to the recent improvements made in machine learning techniques in tuning millions of parameters for the evaluation functions [4]. Some are trying to reproduce the results of AlphaGo (Zero); the current top program uses a combination of alpha–beta search and linear evaluation functions.

We used following six shogi programs for experiments: Apery[1], YaneuraOu[2], Gikou[3], nozomi[4], NNUE[5], and gpsfish[6]. All programs use alpha–beta search with evaluation functions. We performed experiments on a workstation with Intel Xeon CPU E5-2640 2.40 GHz 10core 2CPU and 96 GB memories.

First, we surveyed the strength of those programs, and then, we compared the strength of the proposed method and other systems.

4.1 Tournament Between Single Engines

We performed tournaments between single programs to examine the strength of each single program. Programs have five seconds to search per position. Tournament results between single programs are summarized in Table 1. Winning ratios are calculated from one thousand games, and a draw is counted as 0.5 wins.

From this result, we can see that NNUE is the strongest engine and Apery is the second. All engines are sorted by their strength as NNUE > Apery > Gikou ~ YaneuraOu > nozomi ≫ gpsfish ≫ random. We use Apery as a reference program. Thus, we measure the strength of a system by comparing the winning percentages against Apery. Additionally, we use YaneuraOu as a player in our experiments and YaneuraOu is abbreviated as Yane.

Table 1. Tournament results between single engines against Apery

Engine time	Winning ratio	Win/Draw/Loss		
Yane 5 s	0.3005	290	21	689
Yane 7 s	0.373	363	20	617
Yane 10 s	0.4445	437	15	548
Gikou 5 s	0.2925	288	9	703
nozomi 5 s	0.214	203	22	775
gpsfish 5 s	0.01	6	8	986
NNUE 5 s	0.613	596	34	370

[1] https://github.com/HiraokaTakuya/apery. Accessed May 2019.
[2] https://github.com/yaneurao/YaneuraOu. Accessed May 2019.
[3] https://github.com/gikou-official/Gikou. Accessed May 2019.
[4] https://github.com/saihyou/nozomi. Accessed May 2019.
[5] https://github.com/ynasu87/nnue. Accessed May 2019.
[6] http://gps.tanaka.ecc.u-tokyo.ac.jp/gpsshogi/index.php?GPSFish. Accessed May 2019.

4.2   Tournaments Among Proposed Methods

We performed tournaments among the proposed systems and optimistic voting system against Apery with various advisers, to survey the effect of the strength of the adviser. The results are summarized in Table 2. In the right three columns in the table, each *, **, *** represents $p < 0.05, 0.01, 0.005$, respectively, using chi-squared test with Yates's continuity correction. A hypothesis is that a system is stronger than a single YaneuraOu with 5, 7, and 10 s.

It is clear that the system with the option "Advice" (extended search is limited to the advice) is stronger than the single engines except for random advice. The lowest winning ratio of the systems is 0.376 (adviser : gpsfish), while the winning percentage of the single engines is 0.3005.

Table 2. Tournament results of the proposed methods against Apery

Adviser	Option (or Method)	Winning ratio	Win/Draw/Loss	> YaneuraOu		
				5 s	7 s	10 s
Gikou	Advice	0.483	474 18 508	***	***	
	Both	0.462	449 26 525	***	***	
	All	0.413	400 26 574	***		
	Opt	0.383	381 4 615	***		
nozomi	Advice	0.4215	411 21 568	***	*	
	Both	0.4765	465 23 512	***	***	
	All	0.412	404 16 580	***		
	Opt	0.3775	370 15 615	***		
gpsfish	Advice	0.376	368 16 616	***		
	Both	0.4145	403 23 574	***		
	All	0.394	386 16 598	***		
	Opt	0.1115	110 3 887			
Apery	Advice	0.4885	477 23 500	***	***	
	Both	0.5015	488 27 485	***	***	*
	All	0.4375	422 31 547	***	***	
	Opt	0.427	421 12 567	***	*	
NNUE	Advice	0.523	510 26 464	***	***	***
	Both	0.481	467 28 505	***	***	
	All	0.446	434 24 542	***	***	
	Opt	0.4945	490 9 501	***	***	*
Random	Advice	0.3065	300 13 687			
	Both	0.383	375 16 609	***		
	All	0.4315	420 23 557	***	***	

Option "Advice" is the best for the case if the adviser is stronger or equal to (NNUE, Apery). Option "Both" is the best for the case when the adviser is weaker (nozomi, gpsfish).

If the advice is random, extension always occurs (almost 100%); thus, the consumed time is ten seconds. We can see the performance of random advice with option "All," which is the almost same with that of the single YaneuraOu with ten seconds.

We would like to know the ratios of selection of advice, we summarize the results in Table 3. R1 is a ratio that advice is selected when advice differs from the player's best, R2 is a ratio that advice is selected during all game, and R3 is a ratio that advice differs from the player's best.

Ratios are almost the same between option "Advice" and option "Both". R3 is around 40%, except for gpsfish (50 %), thus, comparison of winning ratio to YaneuraOu with seven seconds is reasonable. The accepted ratio of advice is almost same in Gikou, nozomi, and Apery. R1 of NNUE is higher than the others.

Table 3. Ratios selecting advice during the games

Adviser	Advice			Both			All		
	R1	R2	R3	R1	R2	R3	R1	R2	R3
Gikou	0.226	0.0942	0.416	0.230	0.0955	0.415	0.140	0.0600	0.427
nozomi	0.214	0.0844	0.394	0.246	0.0993	0.403	0.147	0.0596	0.405
gpsfish	0.157	0.0768	0.490	0.169	0.0859	0.509	0.104	0.0516	0.498
Apery	0.221	0.0888	0.402	0.238	0.0962	0.405	0.164	0.0654	0.400
NNUE	0.265	0.100	0.378	0.261	0.105	0.403	0.152	0.0627	0.412
Random	0.0124	0.0118	0.955	0.0113	0.0109	0.960	0.00392	0.00376	0.958

4.3   Tournament Using Two Advisers

We performed tournament experiments with two advisers. We had three game engines in this case; thus, we could apply majority voting. The comparison results are summarized in Table 4.

Table 4. Tournament results against Apery: two advisers

Adviser		Advice	Both	Majority
Gikou	Nozomi	0.521	0.496	0.3865
Nozomi	Gpsfish	0.458	0.4855	0.2815
Gikou	NNUE	0.554	0.532	0.578
NNUE	Gpsfish	0.498	0.5065	0.5255

It can be seen that the wining ratio of the proposed methods increases from the result with a single adviser except for option "Advice" with NNUE and gpsfish.

4.4 Time Extension

Until now, we have performed experiments extending another five seconds (the same with the normal search). We are curious about how performance changes as the extended time differs. Thus, we compared the performance with extending for more or less seconds in this subsection. The results are summarized in Table 5; we used Gikou as adviser and set extending time as half and double in this experiment.

Table 5. Time Extension: Tournament results with adviser Gikou against Apery

Extension	Advice	Both	All
Half	0.4645	0.407	0.3865
Basic	0.483	0.462	0.413
Double	0.494	0.512	0.496

Option "Advice" does not seem to be sensitive to the extension of time as compared to other options. The proposed methods extending ten seconds (Double) are significantly stronger than the single AI with ten seconds, even though the average searching time of the proposed methods is less than ten seconds.

4.5 Discussion

We showed that the proposed method is significantly stronger than the single AIs using the same time in the experiments. Thus, we have shown that the advice can improve the game AIs' strength.

We discuss about the relation between the performance and the adviser's strength. An option "Advice" was better when the advisers were stronger or equal to the player in our experimental results. This indicates that case A is more important than case B in Subsect. 3.1. By contrast, when the advisers were weaker than the player, option "Both" was better. It implies that the importance of case A (advice is better) is low, as discussed in Subsect. 3.1. The advice could be a worse move when the adviser is weak; therefore, searching both moves is required for weaker advisers to avoid selecting the worse advice accidentally.

The majority voting sometimes was stronger than the proposed method; however, the performance degradation was seen in the majority voting and optimistic voting when the system included weaker engine. Our methods were stable compared to the majority voting and optimistic voting.

5   Conclusion

We present methods to enhance search of a game AI using advice from the other game AIs. This research is motivated by the observation that human players can improve their strength with advice. We defined "advice" as the candidate moves selected by adviser and propose the method that make a player search again when the player's move is different from advice.

We performed tournaments among the proposed systems and other methods against a single engine to compare the strength in shogi. We showed the effectiveness of the proposed method from the experimental results and that the game AIs can improve their strength with advice. We additionally found that the advice from weaker game AIs is still useful for a game AI.

In this paper, we simply use the candidate moves as advice. Studying more complex advice such as the candidate moves with their scores and a more complex usage would be interesting. Marcolino et al. studied the diversity of the teams in majority voting [7]; analyzing our methods with the diversity between advisers and players will also be an interesting topic.

References

1. Althöfer, I.: Decision support systems with multiple choice structure. In: Numbers, Information and Complexity, pp. 525–540. Springer, Heidelberg (2000)
2. Althöfer, I., Snatzke, R.G.: Playing games with multiple choice systems. In: International Conference on Computers and Games, pp. 142–153. Springer, Heidelberg (2002)
3. Hassabis, D.: Artificial intelligence: chess match of the century. Nature **544**(7651), 413 (2017)
4. Hoki, K., Kaneko, T.: Large-scale optimization for evaluation functions with minimax search. J. Artif. Intell. Res. **49**(1), 527–568 (2014)
5. Iida, H., Sakuta, M., Rollason, J.: Computer shogi. Artif. Intell. **134**(1–2), 121–144 (2002)
6. Kasparov, G.: The chess master and the computer. New York Rev. Books **57**(2), 16–19 (2010)
7. Marcolino, L.S., Jiang, A.X., Tambe, M.: Multi-agent team formation: diversity beats strength? In: IJCAI 2013, Proceedings of the 23rd International Joint Conference on Artificial Intelligence, Beijing, China, 3–9 August 2013, pp. 279–285 (2013)
8. Obata, T., Sugiyama, T., Hoki, K., Ito, T.: Consultation algorithm for computer shogi: Move decisions by majority. In: International Conference on Computers and Games, pp. 156–165. Springer, Heidelberg (2011)
9. Schaeffer, J.: The games computers (and people) play. Adv. Comput. **50**, 189–266 (2000)
10. Silver, D., Schrittwieser, J., Simonyan, K., Antonoglou, I., Huang, A., Guez, A., Hubert, T., Baker, L., Lai, M., Bolton, A., Chen, Y., Lillicrap, T., Hui, F., Sifre, L., van den Driessche, G., Graepel, T., Hassabis, D.: A general reinforcement learning algorithm that masters chess, shogi, and go through self-play. Science **362**(6419), 1140–1144 (2018)

11. Silver, D., Schrittwieser, J., Simonyan, K., Antonoglou, I., Huang, A., Guez, A., Hubert, T., Baker, L., Lai, M., Bolton, A., Chen, Y., Lillicrap, T., Hui, F., Sifre, L., van den Driessche, G., Graepel, T., Hassabis, D.: Mastering the game of go without human knowledge. Nature **550**, 354–359 (2017)
12. Sugiyama, T., Obata, T., Hoki, K., Ito, T.: Optimistic selection rule better than majority voting system. In: van den Herik, H.J., Iida, H., Plaat, A. (eds.) Computers and Games, Lecture Notes in Computer Science, vol. 6515, pp. 166–175. Springer, Heidelberg (2011)
13. Takeuchi, S.: Weighted majority voting with a heterogeneous system in the game of shogi. In: The 2018 Conference on Technologies and Applications of Artificial Intelligence (TAAI2018), pp. 122–125 (2018)
14. Takizawa, T.: Computer shogi 2012 through 2014. In: Game Programming Workshop 2014, vol. 2014, pp. 1–8 (2014)
15. Taylor, M.E., Carboni, N., Fachantidis, A., Vlahavas, I., Torrey, L.: Reinforcement learning agents providing advice in complex video games. Connect. Sci. **26**(1), 45–63 (2014)

Reducing Partner's Cognitive Load by Estimating the Level of Understanding in the Cooperative Game Hanabi

Eisuke Sato$^{(\boxtimes)}$ and Hirotaka Osawa

University of Tsukuba, Tennnodai, Tsukuba 305-8573, Japan
`hailabsec@iit.tsukuba.ac.jp`

Abstract. Hanabi is a cooperative game for ordering cards through information exchange, and has been studied from various cooperation aspects, such as self-estimation, psychology, and communication theory. Cooperation is achieved in terms of not only increased scores, but also reduced cognitive load for the players. Therefore, while evaluating AI agents playing a cooperative game, evaluation indexes other than scores must be considered. In this study, an agent algorithm was developed that follows the human thought process for guessing the AI's strategy by utilizing the length of thinking time of the human player and changing the estimation reliability, and the influence of this agent on game scores, cognitive load, and human impressions of the agent was investigated. Thus, thinking time was used as an indicator of cognitive load, and the results showed that it is inversely proportional to the confidence of choice. Furthermore, it was found that the mean thinking time of the human player was shortened when the agent used the thinking time of the player, as compared with the estimation of the conventional agent, and this did not affect human impression. There was no significant difference in the achieved score and success rate of the estimation by changing the estimation reliability according to the thinking time. The above results suggest that the agent developed in this study could reduce the cognitive load of the players without influencing performance.

Keywords: Cooperative game · Agent · Hanabi · Human–agent interaction

1 Introduction

The purpose of AI is to cooperate with humans; therefore, an agent utilizing AI needs to act in accordance with the level of human understanding. For example, if an agent uses an unfamiliar technical term while giving instructions, a definition of the term must be included to help the user understand the instructions. This creates an extra cognitive load for the users, although the performance is the same. Cognitive load is one of the indicators of workload estimation that is also used by NASA-TLX [1]. To achieve a cooperative task, it is important to reduce the cognitive load in addition to improving the performance.

© Springer Nature Switzerland AG 2020
T. Cazenave et al. (Eds.): ACG 2019, LNCS 12516, pp. 11–23, 2020.
https://doi.org/10.1007/978-3-030-65883-0_2

The above-mentioned human–AI cooperation is also relevant to games, and Hanabi has particularly been considered as a benchmark for cooperative games [2, 3]. The main feature of this game is that players can see only their opponent's hand, while their own is seen by the opponent; the information provision to the opponent is restricted by tokens. As games are more abstract than real problems, they are easy for AI to work on.

In the past, scores were commonly used as the evaluation criteria for AI agents playing cooperative games. For instance, Osawa developed an AI agent that can infer its state in Hanabi by simulating the viewpoints and actions of Hanabi users. This agent obtained higher scores than agents whose actions were based only on definitive information [4].

However, it is possible that evaluations using only scores cannot sufficiently assess the reduction in the cognitive load. For example, despite a high score, the cognitive load is considered to be high if the human side is required to think for a long time, which is inappropriate for a cooperative AI agent. Therefore, an agent that can reduce the cognitive load while maintaining a reasonable score is required.

In this paper, an agent algorithm is proposed that changes its estimation reliability, an index of how much estimation to use, according to the thinking time of human players to reduce cognitive load. When a player takes a long thinking time twice in a row, the estimation reliability is lowered by increasing the value of the estimation threshold. On the other hand, when a player takes a short thinking time twice in a row, the estimation reliability is increased by decreasing the value of the estimation threshold. The mean thinking time of the player is used as an indicator of the cognitive load. Therefore, the length of the thinking time is important to measure the cognitive load [5]. The score, cognitive load, and evaluation for this agent were investigated by conducting two kinds of experiments while allowing humans to play Hanabi with this agent.

This paper is organized as follows. Section 2 explains the background of Hanabi, and Section 3 describes the difference between the algorithm using the conventional self-estimation strategy and the algorithm developed by this study. In Sections 4 and 5, the two types of evaluation experiment conducted are described. Section 6 presents a discussion of the experimental results, and Section 7 concludes the work.

2   Background of Hanabi

The rules of Hanabi and related works are discussed below.

2.1   Rules of Hanabi

Hanabi is a cooperative card game that can be played by 2 to 5 players; this study deals with 2 players. Hanabi is played with a deck of 50 cards and 8 information tokens. The deck comprises five colors—white, red, blue, yellow, and green—and five numbers, i.e., 1 to 5. There are 10 cards of each color, and there are three cards of No. 1, two each of Nos. 2, 3, and 4, and one card of No. 5. The goal of this game is to complete cards of each color in the ascending order of the numbers from 1 to 5 on the board. Players are dealt five cards each as a hand. The players can only see their opponent's hand and not their own. The remaining cards in the deck are dealt face down. At each turn, a player must choose one of the following three actions: inform the opponent of an attribute of their card, discard a card, or play a card. When a player chooses to inform the opponent of an attribute of their card, the player discloses information regarding the numbers or colors of the cards of the opponent's hand. This action consumes one information token and can be performed only if there are information tokens available. The player must give complete information: for example, if the cards in the opponent's hand are arranged as 1 red, 2 red, 1 white, 2 green, and 1 green and the player wishes to inform the opponent about the instances of red cards, he/she must say, "The first and second cards are red."

When the player chooses to discard a card, one of the cards of his/her hand is discarded, and one card is drawn from the deck to replace it. This action adds one information token. The discarded cards are revealed to both players and cannot be used anymore during the game. When the player chooses to play a card, he/she attempts to use a card from his/her hand to connect to one of the played cards. If the card played is one greater than the number of same colored cards on the board, the play is considered successful, the card is arranged in the corresponding color on the board, and the number is increased by one. However, the card is discarded if the play fails. Whether successful or unsuccessful, the played card is revealed to both players, and the player draws a card from the deck to replace the played card.

- The game ends when any one of the following conditions is fulfilled.
- The players fail to play a card three times.
- Each player performs an action once after the deck becomes empty.
- The players complete up to five of all colors.

The total value of the cards on the board is the final score.

2.2 Works Related to Hanabi

Although studies on agents playing cooperative games have been rare, in recent years, many studies have focused on cooperative card games, such as Hanabi [6–8]. There have been theoretical analyses of Hanabi. For instance, Christopher et al. examined strategies in Hanabi by applying the hat guessing game and showed that players sharing strategies in advance can achieve the highest score with a probability of 75% or more [9]. Similarly, Bouzy developed a best strategy for Hanabi using the hat guessing game and a depth-first search [10]. However, these studies assume that the players share their complicated strategies in advance, which is not suitable for an AI agent cooperating with humans, which is the goal of this research. Nevertheless, there have also been several studies on cooperative behavior in Hanabi. As mentioned earlier, Osawa developed an AI agent that can infer its state in Hanabi by simulating the viewpoints and actions of Hanabi users. He also showed that using this strategy results in higher scores compared with those achieved by agents that act based on only definitive information [4]. Similarly, Eger et al. showed that human beings appreciated the ability of the agent and felt confident when they believed that the agent acted with intention. This was demonstrated by having a human play Hanabi with an agent [11]. Gottwald et al. used an agent that obtains information from human eye movements while playing Hanabi [12]. Furthermore, a strategy for Hanabi using the Monte Carlo tree search has also been developed [13] [14].

However, these studies did not consider human thinking time as a strategy for an agent playing Hanabi, which is an important parameter to measure the understanding of both the human game and agent's strategy.

3 Agent Algorithm Using Thinking Time

An agent that uses the thinking time of the human player to determine his/her understanding of the agent's strategy and changes the estimation reliability was developed.

3.1 Method

This agent changes the degree of confidence threshold in the estimation of the hand according to the length of thinking time of the human player. For example, when the thinking time of the human player is short, the estimation threshold is lowered to increase the reliability, and more estimation results are used. Conversely, when the human player's thinking time is long, the threshold is raised to lower the reliability, and the estimation result is not used much. Furthermore, a short thinking time indicates that the human player is confident of his/her actions; therefore, the human player's actions agree with the model assumed by the agent, resulting in a successful estimation. However, a long thinking time indicates that the human player is not confident of his/her actions, and it is highly probable that the action deviates from the model assumed by the agent, making estimation difficult. Therefore, thinking time is an important measure of human players' understanding of games and opponents. In this paper, two types of agent are discussed: a conventional self-estimating agent that does not use the player's thinking time as a strategy indicator and the proposed self-estimating agent that uses the human player's thinking time as a strategy indicator.

3.2 Conventional Self-Estimation Strategy

An algorithm developed by Kato et al. [15] was used as a conventional self-estimation strategy. The algorithm acts in the following order of precedence.

(A) Play Playable Card Based on Number Information

The agent plays a card whose color is unknown but whose number indicates that it is a playable card. For example, at the start of the game, the No. 1 card becomes a playable card regardless of the color.

(B) Inform Playable Card

If the player has a playable card, but incomplete information, the agent informs him/her; if both the number and color are unknown to the human player, the agent presents either information at random.

(C) Play Playable Cards Based on Self-Estimation

The agent uses a simulation of the human player's observation to estimate the agent's own hand. The agent considers the set of possible hands at each time and sorts the elements in this set by the number of occurrences; when the value of the highest occurrence number is a constant multiple or more of the value of the next-highest occurrence number, this card is estimated to be equal to the value of the highest occurrence number. The agent plays if this value is equal to the playable card.

(D) Inform a Discardable Card

If the human player has a discardable card and can understand that it is discardable by revealing the number or color, the agent presents the information.

(E) Play Playable Cards Based on Board

If the player's hand and the observable card on the board show that a playable card is present in the agent's hand, the agent plays that card.

(F) Discard a Discardable Card

If the player's hand and the observable card on the board show that a discardable card is present in the agent's hand, the agent discards that card.

(G) Discard a Random Card

The agent discards the card with the least information in its hand. If there are multiple such cards, the agent discards a random card among these.

3.3 Self-Estimation Strategy That Changes Estimation Reliability According to Thinking Time

We modified Kato's self-estimation strategy by adding the algorithm that changed the estimation reliability according to the thinking time of the human player. As a result, the agent can be expected to act according to the human player's understanding of the game or the agent's strategy.

First, the agent records the time taken by the human player to choose an action for the previous five turns. From this, the agent derives the mean (M) and standard deviation (SD) of the thinking time. If the next thinking time is more than the value of the M + SD, it is considered as a long thinking time, and, if it is less than the value of M − SD, it is considered as a short thinking time. Depending on the deviation of the thinking time for the previous five turns, the standard deviation may be larger than the mean, and it is difficult to define a time as a prompt decision. Therefore, if the human player takes a thinking time of 4.1 s or less, it is defined as a short thinking time regardless of the mean or standard deviation. This is because 4.1 s is 1 standard deviation less than the mean of the mean thinking time of each player in the Hanabi experiment previously performed by Kato on 12 university students [15]. This value is the mean of the threshold values by which each player is determined to have made a prompt decision. Therefore, it is considered to be an appropriate threshold to determine a short thinking time. The following hypotheses were made regarding the length of the thinking time.

- If the player takes a long thinking time, he/she is not confident in his/her action.
- If the player takes a short thinking time, he/she is confident in his/her action.

In other words, it was hypothesized that the thinking time and confidence in selection are inversely proportional to each other. So, when the player continues to think for a long time before selecting an action, indicating that he/she is not confident in the selection, it is considered that the agent's strategy is not well understood. Hence, the probability of failure in estimation is higher than usual. This is because the agent's self-estimation assumes that the human player acts in the same thinking pattern as the agent, and it is essential that the human player understands the agent's strategy to some extent. Thus, as discussed in Sect. 3.2 (C), this agent follows the estimation if the value of the most likely card in the estimation is a constant multiple or more of the value of the next-highest occurrence number. Therefore, when the player takes a long thinking time twice in a row, the estimation reliability is lowered by increasing the value of this constant.On the other hand, when the human player continues to take a short thinking time in selecting the action, he/she is considered to have confidence in his/her choice. However, this situation can result from two circumstances. The first is that the agent's strategy is correctly understood, while the second is that the agent's strategy is misinterpreted, and the human player's action is based on that misinterpretation. In the first case, the agent should increase the confidence of the estimate, while, in the second case, the agent should reduce the confidence of the estimate. We used the choice of the human player's action to determine the situation. Specifically, when the human player took a short thinking time twice in a row, the estimate reliability was increased, but if his/her card failed to play in this situation, the estimate reliability was lowered.

4 Thinking-Time Experiment

Two experiments were performed in this study: a thinking-time experiment and a cognitive-load experiment. The objectives of the thinking-time experiment were:

A)To confirm the validity of the hypotheses about thinking time and confidence of choice described in Sect. 3.3.

B)To compare the performance of an agent that uses the human player's thinking time with that of the conventional agent, and to determine the parameters that affect the human impression on the agent.

4.1 Interface

An interface implemented by Kato et al. using Visual C++ was used to realize a game between a human and an AI agent [15], which is shown in Fig. 1.

The user's hand is displayed at the lower left of the screen, and the opponent's hand is displayed at the upper left of the screen. The upper center represents the current board, and the numbers from left to right in the lower center represent the number of cards remaining in the deck, the number of information tokens, the number of failed plays, and the score, respectively. On the right, the cards that have been played, failed, or discarded are displayed.

4.2 Evaluation Method of the Thinking-Time Experiment

In the first thinking-time experiment, to check the impression of the users on the agent, the game was recorded, and an impression evaluation questionnaire was conducted after the experiment. The following questionnaire was used for the evaluation.

Q1. Did you feel that the agent has the intention?
Q2. Did the agent get used to the game?
Q3. Was the agent friendly?
Q4. Did the agent understand you?
Q5. Was the agent wise?
Q6. Did the agent act as you wanted?
Q7. Did you understand the intention of the agent's action?
Q8. Do you think that the agent's behavior was consistent?
Q9. Did you try to increase the score?
Q10. Was the game easy to play?
Q11. Do you consider any of your own actions unsuccessful?
Q12. Were there any unreliable actions, whether they were correct or not?
Q13. Did you go ahead with the game?
Q14. Were you satisfied with the outcome of the game?
Q15. Do you have any comments? (Free description)

Q1 to Q5 were based on The Godspeed Questionnaire Series by Bartneck et al., which is one of the indexes used for evaluation [16]. All questions except Q15 used the 7-step Likert scale for the responses, where 0 means "No" and 7 means "Yes".

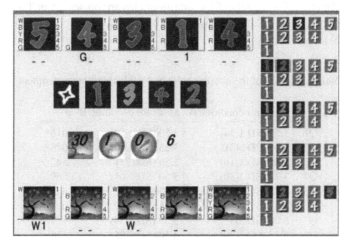

Fig. 1. Interface of Hanabi

4.3 Procedure of the Thinking-Time Experiment

Four male college students in their twenties participated in the experiment. The participants played the game twice after receiving instructions about Hanabi's rules and interface. Each participant played two games, one with the conventional self-estimation agent (condition A) and the other with the proposed self-estimated agent (condition B).

Therefore, four samples were gathered for each condition. During the game, the participants were asked to record their confidence in their choices of actions after each turn. Two types of decks were prepared and used in the experiment. The questionnaire was conducted at the end of each game.

4.4 Result of the Thinking-Time Experiment

The mean thinking time for the turns where the participants were confident in their actions was 21.1 s (SD: 15.1), whereas that for the turns where they were not confident was 39.8 s (SD: 17.8). Furthermore, the mean thinking time for the entire game was 24.7 s (SD: 17.7). The results of the t-test (p < .05) showed a significant difference between the mean thinking time for the turns where the participants were confident and the mean thinking time (p < .05) for the entire game. Similarly, there was a significant difference between the mean thinking time of the turns where the participants were not confident and the mean thinking time (p < .01) for the entire game. The mean score for condition A was 17.5 points (SD: 5.4), and that for condition B was 17.8 points (SD: 2.9); the results of the t-test showed no significant difference. Under condition A, cards were estimated 7.5 times per game (SD: 5.7) on average, and the mean success rate was .57 (SD: .34). Under condition B, cards were estimated 6.5 times per game (SD: 2.2) on average, and the mean success rate was .57 (SD: .17); the results of the t-test once again showed no significant difference. The results of the questionnaire are presented in Table 1. It was observed that, for Q1 and Q2, there was no significant difference between conditions A and B (p < .05).

Table 1. Results of the questionnaire in the thinking-time experiment

	Mean for condition A	Mean for condition B	p
Q1	5.50 (SD 1.34)	3.50 (SD 1.84)	.016*
Q2	6.25 (SD .43)	3.25 (SD 1.60)	.046*
Q3	3.25 (SD 1.16)	2.50 (SD .77)	.547
Q4	3.75 (SD 1.30)	3.50 (SD 1.61)	.889
Q5	3.75 (SD .97)	2.75 (SD 1.83)	.474
Q6	3.25 (SD 1.79)	2.25 (SD 1.47)	.613
Q7	6.00 (SD .63)	3.50 (SD 1.34)	.127
Q8	5.25 (SD 1.30)	4.50 (SD 2.05)	.638
Q9	6.25 (SD .74)	5.75 (SD 1.16)	.664
Q10	5.25 (SD 2.05)	3.75 (SD 2.04)	.576
Q11	5.25 (SD 1.83)	6.50 (SD .77)	.391
Q12	5.75 (SD 1.09)	6.00 (SD .89)	.638
Q13	5.00 (SD 1.67)	4.50 (SD 1.34)	.804
Q14	4.00 (SD 2.55)	2.50 (SD 1.84)	.547

*p < .05

5   Cognitive-Load Experiment

Because there were several questionnaire items in the thinking-time experiment, there was a possibility of pseudosignificant differences. Therefore, the questionnaire items for the cognitive-load experiment were reduced. The following were the objectives of the cognitive-load experiment:

A) To compare the cognitive load imposed by the two agents.

B) To observe the significant differences in the questionnaire items that showed changes in the thinking-time experiment.

C) To observe the significant differences in the scores and estimates when the sample size was increased.

5.1   Evaluation of the Cognitive-Load Experiment

To observe the changes in the cognitive load due to changes in agents, the mean thinking time of the human player per turn was recorded. To check the impression on the agent, the game was recorded and impression evaluation questionnaires were conducted after the experiment. The following are the contents of the questionnaire.

Q1. Did you feel that the agent has the intention?

Q2. Did the agent get used to the game?

Q3. Did you understand the intention of the agent's action?

Q4. Do you think that the agent's behavior was consistent?

Q5. Was the game easy to play?

Q6. Were you satisfied with the outcome of the game?

Q7. Do you have any comments? (Free description)

From the first experiment, it was observed that significant differences appeared in Q1 and Q2 during the experiment, and it was speculated that Q3 and Q4 played an important role in determining if humans recognized the difference between the two agents. In this experiment, the questions were reduced to six; the last two questions about the game were added to the existing four questions. All questions except Q7 used the 7-step Likert scale for the responses.

5.2   Procedure of the Cognitive-Load Experiment

The participants included 12 male and 8 female university students. None of the participants knew the rules of Hanabi. Each participant played the game twice after receiving instructions about Hanabi's rules and interface. The first game involved the conventional self-estimation agent (condition A), and the second involved the proposed self-estimated agent (condition B). The participants then changed the order and played the same game one by one. Therefore, two samples were collected per participant for each condition. Four types of deck were prepared and used for the experiment. A questionnaire was conducted at the end of each game. The order of conditions A and B was changed for each participant for counterbalance.

5.3 Results of the Cognitive-Load Experiment

In this experiment, the game was played 40 times for each condition, i.e., A and B. Therefore, there were 40 pieces of data each for conditions A and B. The mean score for condition A was 16.6 points (SD: 4.6), whereas that for condition B was 17.1 points (SD: 4.1); the results of the t-test showed no significant difference. Under condition A, cards were estimated 5.1 times per game (SD: 3.6) on average, and the mean success rate was .66 (SD: .27). Under condition B, cards were estimated 6.1 times per game (SD: 4.3) on average, and the mean success rate was .71 (SD: .27). The results of the t-test showed no significant difference. The results of the questionnaire are presented in Table 2; no significant difference was observed in the results. The mean thinking time in one turn of condition A was 16.8 s (SD: 7.11) and that in one turn of condition B was 15.12 s (SD: 4.76); the results of the t-test showed a significant trend ($+p < .10$).

Table 2. Results of the questionnaire in the cognitive-load experiment

	Mean for condition A	Mean for condition B	P
Q1	4.78 (SD 1.87)	4.85 (SD 1.79)	.395
Q2	4.78 (SD 1.48)	4.67 (SD 1.76)	.440
Q3	4.88 (SD 1.64)	5.18 (SD 1.67)	.314
Q4	4.60 (SD 1.61)	4.91 (SD 1.70)	.232
Q5	4.78 (SD 1.58)	4.82 (SD 1.98)	.405
Q6	4.13 (SD 1.56)	4.15 (SD 1.70)	.480

6 Discussion

The discussion of these experiments is shown below.

6.1 Discussion of the Thinking-Time Experiment

In the thinking-time experiment, there was a significant difference between the mean thinking time of the turns where the participants were confident and the mean thinking time of the entire game; similarly, there was a significant difference between the mean thinking time of the turns where participants were not confident and the mean thinking time for the entire game. Therefore, the hypotheses that thinking time and confidence in selection are inversely proportional was validated. In this experiment, no significant difference was observed in the scores, the number of estimations, and the success rates between the two conditions. Therefore, it was concluded that agents that used the thinking time had no influence on the score in this experiment. Furthermore, the results of Q1 and Q2 in the questionnaire in Table 1, although based on a small sample, show that when the estimation reliability is affected by the thinking time, the users may feel that the agent is either unwilling or not accustomed to the game. This could be because, when the estimation reliability decreases because of long thought times among all the participants, the agent acts with greater emphasis on the definite information. To the human user, it may appear that the agent is not used to the game.

6.2 Discussion of the Cognitive-Load Experiment

In the cognitive-load experiment, there was a significant trend in the mean thinking time between conditions A and B, which shows that agents using thinking time may have a smaller mean thinking time than the conventional agents do. This could be because the use of thinking time allows greater human–agent cooperation, which may reduce the cognitive load. As no significant difference was observed between the scores obtained for conditions A and B, similar to the thinking-time experiment, it was concluded that the proposed agent algorithm had no influence on the score. This could be because the mean score of the participants was high while playing with the conventional agent, and there was no further increase in the score when the agent was changed. In fact, the mean scores for both conditions A and B were 16 or more, and, according to the rule book, this is regarded as an excellent score [17]. Similarly, in the first experiment, there was no significant difference between the two conditions in the number of estimations and the success rate. Therefore, it was concluded that the proposed agent algorithm had no influence on the number of estimations in this experiment. Furthermore, Tables 1 and 2 show no significant difference between the questionnaire results of conditions A and B. Therefore, the effect of the proposed agent algorithm on the impression of people was not observed clearly. This result is different from the questionnaire result of the first experiment; this could be because of a bias in the data owing to the small sample size.

6.3 Limitations and Future Work

This experiment showed that changes in the agent algorithms may have reduced the cognitive load. However, as the questionnaire results showed no significant differences between conditions A and B, the player of the game was likely to be unaware of the change in agent. Therefore, we believe that it is not possible to improve the evaluation of humans by using the agent algorithm developed in this study. So, it would be necessary to modify the agent's strategy so that it is more easily recognized by humans without raising the cognitive load. For example, in addition to changing the estimation reliability of the agent, adding an observable change, such as the time taken to select the agent's action, can solve this problem. This study shows that agents that use thinking time may reduce the cognitive load of their partners. Using these results, development of more-cooperative agents can be expected. For example, an agent that can reduce the thought process of the user by changing his/her speech or action based on the user's cognitive ability could be developed.

7 Conclusion

In this study, an agent algorithm was developed that uses a human player's thinking time to estimate the degree of confidence in choosing an action and uses this as an index to change the estimation reliability. Furthermore, the changes in the score, thinking time, and impression of the human users were compared with those obtained using the conventional self-estimation strategy. The results validated the hypothesis that the human player's thinking time is inversely proportional to the selection confidence. Additionally, it was observed that the proposed agent algorithm could reduce the cognitive load of the human user. This can be a new metric that can replace the score obtained by a cooperative agent. However, the results of the questionnaire were not clear regarding the effect on the human impression of the agent. A future improvement would be the development of agents that use the thinking time more appropriately. This improvement is expected to be achieved through more complicated hypotheses on human thinking time that are closer to the actual human thought process.

Acknowledgement. This research was supported by JSPS Research Grants JP26118006, JP18KT0029.

References

1. Hart, S.G.: Nasa-task load index (NASA-TLX); 20 years later. Proc. Hum. Fact. Ergon. Soc. Annual Meet. **50**(9), 904–908 (2006)
2. Jaderberg, M., Czarnecki, W.M., Dunning, I., Marris, L., Lever, G., Castaneda, A.G., Beattie, C., Rabinowitz, N.C., Morcos, A.S., Ruderman, A., Sonnerat, N., Green, T., Deason, L., Leibo, J.Z., Silver, D., Hassabis, D., Kavukcuoglu, K., Graepel, T.:Human-level performance in first-person multiplayer games with population-based deep reinforcement learning (2018). arXiv:1807.01281
3. Venture Beat, Google Brain and DeepMind researchers release AI benchmark based on card game Hanabi.https://venturebeat.com/2019/02/04/google-brain-and-deepmind-researchers-release-ai-benchmark-based-on-card-game-hanabi/(accessed 2019–05–03).
4. Osawa, H.: Solving Hanabi : estimating hands by opponent's actions in cooperative game with incomplete information. In: AAAI Workshop, Computer Poker and Imperfect Information, pp. 37–43 (2015)
5. Khawaja, M.A., Ruiz, N., Chen, F.: Think before you talk: an empirical study of relationship between speech pauses and cognitive load. In: Proceedings of the 20th Australasian Conference on Computer-Human Interaction: Designing for Habitus and Habitat (OZCHI 2008), pp. 335–338 (2008)
6. Bard, N., Foerster, J.N., Chandar, S., Burch, N., Lanctot, M., Song, H.F., Parisotto, E., Dumoulin, V., Moitra, S., Hughes, E., Dunning, I., Mourad, S., Larochelle, H., Bellemare, M.G., Bowling, M.: The Hanabi Challenge: A New Frontier for AI Research (2019). arXiv: 1902.00506
7. Canaan, R., Shen, H., Torrado, R., Togelius, J., Nealen, A., Menzel, S.: Evolving agents for the Hanabi 2018 CIG competition. In: 2018 IEEE Conference on Computational Intelligence and Games (CIG), pp. 1–8 (2018)
8. van den Bergh, M., Spieksma, F., Kosters, W. Hanabi, a co-operative game of fireworks. Leiden University, Bachelor thesis (2015)
9. Cox, C., De Silva, J., Deorsey, P., Kenter, F.H.J., Retter, T., Tobin, J.: How to make the perfect fireworks display: two strategies for Hanabi. Math. Mag. **88**, 323–336 (2015)
10. Bruno, B.: Playing Hanabi near-optimally. In: ACG 2017: Advances in Computer Games, pp. 51–62 (2017)
11. Eger, M., Martens, C., Córdoba, M.A.: An intentional AI for Hanabi. In: IEEE Conference on Computational Intelligence and Games (CIG), pp. 68–75 (2017)
12. Gottwald, E.T., Eger, M., Martens, C.: I see what you see: integrating eye tracking into Hanabi playing agents. In: Proceedings of the AIIDE workshop on Experimental AI in Games (2018)

13. van den Bergh, M.J.H., Hommelberg, A., Kosters, W.A.: Aspects of the cooperative card game Hanabi. In: BNAIC 2016: Artificial Intelligence, pp. 93–105 (2016)
14. Walton-Rivers, J., Williams, P.R., Bartle, R., Perez-Liebana, D., Lucas, S.M.: Evaluating and modelling Hanabi-playing agents. In: 2017 IEEE Congress on Evolutionary Computation, pp. 1382–1389 (2017)
15. Kato, T., Osawa, H.: I know you better than yourself:estimation of blind self improves acceptance for an agent. In: HAI 2018 Proceedings of the 6th International Conference on Human-Agent Interaction, pp. 144–152 (2018)
16. Bartneck, C., Kulić, D., Croft, E., Zoghbi, S.: Measurement instruments for the anthropomorphism, animacy, likeability, perceived intelligence, and perceived safety of robots. Int. J. Soc. Rob. 1(1), 71–81 (2018)
17. Ultra Board Games,"Hanabi Game Rules". https://www.ultraboardgames.com/hanabi/game-rules.php. Accessed 03 May 2019

Making a Better Game: The History of Cluster

Gregory Schmidt[1] and Philip Shoptaugh[2(⊠)]

[1] Rockwell Automation, Mayfield Heights, OH, USA
gschmidt958@yahoo.com
[2] Shoptaugh Games, Oakland, CA, USA
philip@shoptaugh.com

Abstract. The authors present a case study of the initial inspiration and design process that led to successfully optimized versions of the game "Cluster". Various aspects of game design are examined in the context of human and computer assisted playtesting.

1 Introduction

Cluster is a two player connection game designed by Philip Shoptaugh in 1972. This article describes the initial inspiration and development of successively optimized versions of the game. In doing so, the primary question addressed by this work is: "How can playtesting be effectively applied to successively improve the design of a game?" Herein, the following aspects of the game's design are addressed:

- Two player games are sometimes produced via the recombination of elements obtained from the domain of existing games [3]. An example is given whereby a novel "hybrid" two-player game is derived from a cross between a two player game and a single player puzzle.
- The optimization of an existing game is demonstrated via applying design heuristics gleaned from insights acquired through playtesting.
- The utilization of computer assisted game design software is examined. Prior work in this area has been primarily directed towards the goal of automatic invention of new games via automated recombination and evaluation of existing two player games [2, 3]. These systems typically employ a generic AI. This work illustrates utilizing a game design system that incorporates a high quality custom AI specifically designed to play variations of a single game. The rule set is fixed, whereas the board topology and initial piece type distribute and initial placement is allowed to vary. This work is aimed at increasing the interaction between the game designer and the game design software by enabling the designer to quickly generate and playtest proposed variations.

The resulting design rules and playtesting techniques, shown to be successful for Cluster, should yield applicability to a larger domain of games. The applicable classes of games include connection and territorial games, where the primary design goal is to optimize the board topology, the distribution of piece types, and the initial piece placements.

T. Cazenave et al. (Eds.): ACG 2019, LNCS 12516, pp. 24–40, 2020.
https://doi.org/10.1007/978-3-030-65883-0_3

1.1 Rules of Cluster

The Cluster game board consists of a pattern of holes, some which are deep (as identified with a chamfer), and some which are shallow (without the chamfer). Each player, identified as black or white, has two rows of pegs of their respective color inserted into holes. Some pegs are tall and some are short. Based on the combination of peg length and hole depth, three levels are possible, short, medium, and tall which can also be identified numerically by the numbers 1, 2, and 3 respectively (Fig. 1). With white opening, the players take turns moving their pegs until the winning player forms a contiguous group of pegs of their respective color and all pegs of the group are at medium height (level 2). Pegs can move in two ways. They can either step to an adjacent empty hole or they can jump any number of pieces on the lateral or on the diagonal, regardless of ownership or height of the pieces being jumped, then landing in the first vacant hole.

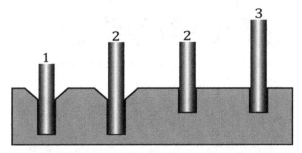

Fig. 1. Peg height as determined by peg size and hole depth.

2   The Genesis of a New Game

Interestingly, the inspiration for Cluster originated from combining elements from two sources, Fig. 2, the "Lines of Action" (LOA) game [1], and Fig. 3, the "Plunging Pegs" (PP) puzzle. In LOA, the goal is to form a connected group of one's own pieces. In PP, the goal is to align pegs of varying lengths into holes of varying depths. This ancestry highlights an important aspect of game design whereby new, "hybrid" games can emerge via recombination of elements from existing games [3], and in this case, includes the puzzle domain. It is a process somewhat akin to that of genetic recombination.

2.1 Cluster's Lineage

During Cluster's inception, Shoptaugh was working with a company named Four Generations in Sebastopol California. One of the products produced was a puzzle called "Plunging Pegs". It is a puzzle made out of a single block of wood with four holes, each at a different depth. It has eight pegs of differing lengths. The object of the puzzle is to stack two pegs into each hole such that the tops of the extending pegs are all at the same level. After playing with the puzzle quite a bit, Shoptaugh thought that it would be fun

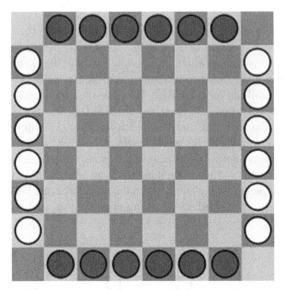

Fig. 2. Lines of Action game starting position.

Fig. 3. Plunging Pegs Puzzle.

to make a two player strategy game using the same "leveling" concept. At the time he had just created a couple of other games for Four Generations, a game called TAU (now called Calypso) and another game called Impasse (now called Shuttles). Shoptaugh also happened to be reading Sid Sackson's newly published book, "A Gamut of Games", and had reviewed Claude Soucie's game "Lines of Action" with its "grouping concept". In a moment of inspiration, Cluster was born as a two player game with two different depths of holes, and two different lengths of pieces. The object of the game is to arrange all of one's pieces clustered together in any free formed group, all the same level, anywhere on the game board. Shoptaugh later met collectively with Sid Sackson, and Claude Soucie. Cluster met with Soucie's approval as he was complimentary of the game[1].

3   From Concept to Realization

There is no doubt that settling on the general form of the rules is a crucial milestone in a game's development. However, in addition to the rules, one must also consider the actual topology of the board and pieces. For Cluster, this topology is represented both by the number and arrangement of shallow and deep holes, along with the number and initial placement of the black and white pegs. These variables represent degrees of freedom which must ultimately be finalized prior to claiming completion of the game design. In the case of Cluster, handmade wooden prototypes were tediously produced for the purpose of experimenting with various board configurations, various depths of holes, differing initial peg positions, and even alterations to the initially proposed rules.

3.1 Design Considerations

One initial design goal was to require each player to move each one of their pieces at least once during the course of the game. This requirement was met by ensuring that none of the initial peg placements were at level 2 (Fig. 4). Specifically, a player's tall pieces are placed along the furthest row consisting of shallow holes (at opposite sides for each player). The short pieces are placed at the second furthest row of deep holes (and on the opposite side of the board with respect to that player's tall pieces).

This arrangement has the additional advantage of promoting the strategic interaction of both players' pieces during the early phase of the game since there are initially six jump moves available to each player. The remaining holes in the center of the board are spaced so that there would be both shallow and deep holes in the center of the game board, with an equal number of each kind. In order to provide a visual cue of the differing hole depths, a chamfer (i.e. countersink) appears on the top of each deep hole. This distinction reduces the memory burden of the two players allowing increased focus on strategic and tactical concerns thereby enhancing the clarity of the game[2].

Initially, the pegs were allowed to jump over other pegs only if the levels of all the pegs along the line to be jumped over were either less than or equal to the level of the jumping peg. This design proved to be both overly confusing and restrictive and thus was abandoned (Fig. 5). The final rules are both simple and elegant. A player can either step to an adjacent empty hole or jump any number of adjacent pieces on the lateral or on the diagonal, regardless of ownership or height of the pieces being jumped, landing in the first vacant hole. The layout of the shallow and deep holes in the "honeycomb" pattern was inspired by Hex, although maintaining symmetry and equal distribution of the two depths were guiding principles.

---

[1] Personal communications with Shoptaugh in 2009.

[2] http://www.thegamesjournal.com/articles/DefiningtheAbstract.shtml.

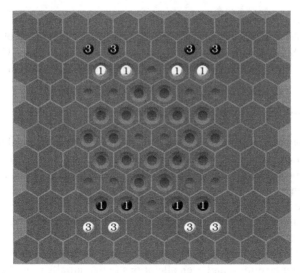

Fig. 4. Starting configuration. All pegs must be moved to win the game since none are initially at level 2.

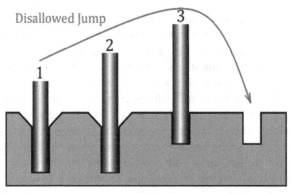

Fig. 5. Initially jumps could only occur over pegs at a less than or equal length of the jumping peg.

During the course of experimenting with the game play, there was a desire to avoid the potential problem of "first player advantage". Fortunately, this was not an issue, because in many cases the second player can jump over the opponent's initially moved piece, thereby advancing further (Fig. 6). Additionally, the layout of the differing hole depths is such that neither player can achieve an insurmountable blocking configuration during the early stages of the game. In order for a player to build a "wall", both tall and short pieces must work in unison to form a string of level 2 pieces. Even if a player creates a wall, the opponent may be able to overcome it by jumping over the wall, and in some cases strategically jump from one end of the game board to the opposite end.

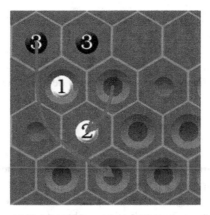

Fig. 6. Following white's initial move, black is able to advance over two pieces.

Two critical elements in winning the game are timing and position. Games between players of similar strength are usually very close with only one or two moves apart from achieving the winning "cluster" formation. As in any good strategy game, it is necessary to think ahead and carefully observe the consequences of the opponent's moves. The rules for Cluster are simple, but players of the game develop complex multi-move strategies that emerge from these simple rules.

3.2 Initial Release

The initial version of the game contained 46 holes (referred to generically as "Cluster-46") with 20 shallow holes and 26 deep holes, and with the initial peg location as shown in Fig. 7. Although two larger prototypes were also proposed, due to cost and manufacturing constraints, the smaller version was selected for production.

Fig. 7. The original Cluster game with Four Generations packaging, circa 1973.

Four Generations made the Cluster game for several years, circa 1972–1975, before the company went bankrupt. The combination of the peg movement rules and the connection goal along with the additional constraint of leveled pieces make Cluster a novel enough game to warrant a patent. The game was patented in 1974 [4] and the patent lists several variations, including holes which have three depths rather than two.

4 Cluster-64

During 2009, Schmidt corresponded with Shoptaugh to create an Axiom [5] computer version of the original game. At that time, he had expressed belief that the initial design of the game could be improved and wanted to experiment with some ideas. However, testing new variations of the game would require the effort intensive work of creating physical wooden prototypes. Obviously, the number of prototypes created this way is limited by both patience and physical resources.

4.1 Automating the Prototype Generation Process

What if instead of creating a Cluster computer game that was bound to a specific configuration, the designer could instead design their own Cluster game and playtest it? This idea formed the basis of a new program called "Cluster Designer"[3]. When invoked, Cluster Designer initially presents an empty board void of all holes. The game designer can subsequently place holes, both shallow and deep, on the board to create a unique hole pattern. Once the hole pattern is fully specified, an arbitrary number of black and white pegs can then be placed to form the initial placement of pieces. The completed game variation can then be saved to a computer file. The game variant is now ready to be playtested. Prior to playing a game, black and white can be assigned to either a human or an AI player.

A Cluster variation can alter the number, layout, and distribution of shallow and deep holes, as well as the number, initial placement, and distribution of short and tall pegs. Otherwise, all variations share the same rules and end of game condition. Cluster Designer's AI was designed specifically to play these variations generically. For example, the AI examines features which are common to all variations such as encouraging level 2 piece groupings and peg mobility while discouraging other negative features such as isolated pegs.

Cluster Designer was implemented as an Axiom game and presented to Shoptaugh who then began creating experimental variations of Cluster. After playtesting a variety of games ideas, he settled on a 64 hole configuration (Fig. 8), and via additional playtesting, concluded that Cluster-64 is superior to the original Cluster-46. The number of pieces per player was increased from 8 to 12 as it was determined that 12 pieces led to deeper game play with a more satisfying tempo. Having 12 pieces increases the challenge of timing the coordination of moves required to bring all pieces into play. The board size increased as well since increasing the number of pieces naturally led to a corresponding increase in the number of holes, equally split between shallow and deep.

---

[3] Understandably, due to a desire to prevent the proliferation of endless variations of Cluster, Shoptaugh requested that Cluster Designer not be made publically available.

Fig. 8. Cluster-64, a.k.a. "Cluster Tournament".

The revised game was not marketed although a small number of handmade copies were produced by Shoptaugh in his workshop. However, Cluster-64 was made more widely available via a subsequent Axiom program that includes the new Cluster-64 design along with the original Cluster-46 (a.k.a. "Cluster Classic") version (Fig. 9). Both stand-alone Axiom PC version[4] and Zillions of Games™[5] versions are available.

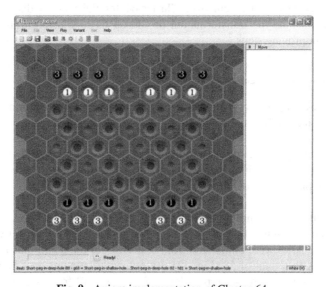

Fig. 9. Axiom implementation of Cluster-64.

[4] http://www.boardgamegeek.com/filepage/46261/cluster-axiom-computer-game-pcs

[5] http://www.zillions-of-games.com/cgi-bin/zilligames/submissions.cgi?do=show;id=1760

5   Cluster-58

Fast forwarding to 2015, Shoptaugh discovered a third incarnation of Cluster (Fig. 10), played on a 58 hole board (30 shallow, 28 deep).

Fig. 10. The new Cluster-58.

Cluster-58 is an improvement over Cluster-64 for the reasons discussed below. Here we will illustrate the transformation of Cluster-64 to Cluster-58 as a series of incrementally improving steps.

5.1   The Refinement Process

First, the blank area gap (no holes) in the two starting rows were eliminated and the remaining six shallow holes were brought together to form a single contiguous row (Fig. 11). By eliminating the gap, players can now jump laterally across the back row. It also has the benefit of increasing clarity by eliminating potential confusion as to whether or not a player is allowed to jump laterally over the gap. Furthermore, it improves the end game play, as a player can now jump a piece laterally across the entire row, unrestricted by the former gap.

Secondly, after observing the use of the side holes during play of many games, it was determined that some of the outside, deep holes, were very infrequently used, so a total of six deep holes were eliminated from both sides of the board thus reducing the number of holes to a total of 58 (Fig. 12). By reducing the number of holes from 64 to 58, and adding one more piece per player (from 12 to 13 for a total of 26 pieces), there is improved interaction and increased competition for critical holes between the two players. However, removing these six deep holes left some undesirable gaps in the middle row.

Thirdly, some of the holes were then rearranged in such a way as to both remove the gaps and to achieve a more even distribution between the two hole types (Fig. 13).

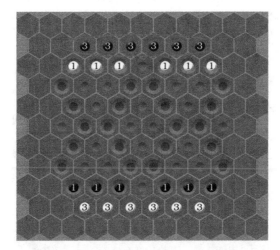

Fig. 11. Cluster-64 with starting row gaps removed.

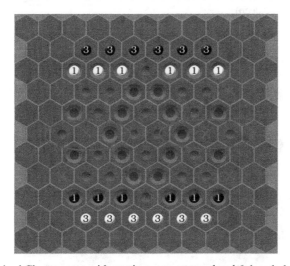

Fig. 12. Revised Cluster game with starting gaps removed and 6 deep holes eliminated.

Finally, since there is no way to divide 58 in half (yielding 29 holes of each depth) while maintaining a symmetrical board, the 2nd and 8th row shallow holes were changed to deep holes resulting in 30 shallow holes and 28 deep holes. Also, an additional piece was added for each player in order to increase the interaction between players (e.g. vying for the same shallow hole at the end of the game) as well as for aesthetic reasons (Fig. 14).

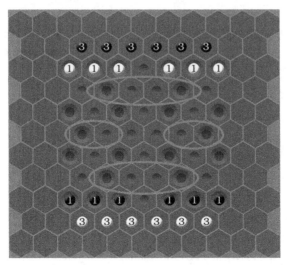

Fig. 13. Revised Cluster game following hole re-arrangement. Note the large number of shallow holes (32 shallow vs. 26 deep).

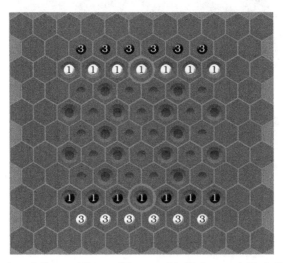

Fig. 14. Final version of Cluster (Cluster-58) now with 28 deep and 30 shallow holes and two additional pieces.

5.2 The Finalized Design

Visually, the new shape with its slightly truncated corners, (due to the gap elimination), is more aesthetically pleasing to the eye.

The revised game board is now slightly longer than it is wide and simply "felt" right to the designer. Most games end with a clear winner and rarely does a situation occur where neither player can force a win thereby ending the game in a draw. Although the game works in a variety of configurations, it is now finalized in its preferred configuration as Cluster-58.[6]

6 Cluster Strategy

Cluster is a game where "efficiency" is of key importance. Quite frequently the games are won or lost by just a few moves, so players must be careful not to lose tempo by playing subpar moves. Listed here are a few important strategic and tactical concepts which are intrinsic to Cluster game play.

Definitions:

- A "group" is defined as a collection of connected pieces, of uniform color, all at height 2.
- A "cluster" is defined as a group containing all 13 pieces of a single color.
- A "liberty" is defined as an empty hole that is immediately adjacent to (i.e. "touching") one's group.
- A "sentinel" is defined as a piece strategically placed for the purpose of inhibiting the further growth and/or eventual completion of the opponent's cluster.

Strategic and tactical concepts:

1. **Mobility** – Maximizing the number of moves available to one's own pieces while minimizing the number of opponent moves.
2. **Center control** – Frequently, winning clusters occur at the center of the board so it's often advantageous to occupy centrally located spaces.
3. **Advancement** – It is important in advancing one's pieces to desired spaces quickly by leveraging jump moves.
4. **Group size** – In many cases, it is good to favor moves which increase the size of one's largest group.
5. **Wall formation** – A player forms a connected string of pieces, (i.e. a "wall") which splits the opponent's pieces into two groups thereby making it harder for them to unite as a single cluster.
6. **Fork** – Moving a piece to an intermediate location such that in a subsequent move, it can connect to a group in more than one way.
7. **Adequate Liberties** – One must ensure that sufficient liberties exist in order to complete the cluster. Note that this includes taking into consideration the hole depths of these liberties in conjunction with the length of the remaining "stray" pieces such that they will eventually mate at the correct height as they are assimilated into the group.
8. **Offensive moves** – A player can inhibit the formation of the opponent's cluster by deliberately placing a piece in one of the opponent's liberties. By strategically placing a sentinel, it's possible to "starve" the opponent's group thereby thwarting

---

[6] Cluster-58 will be commercially marketed in late 2016.

its completion. If the sentinel is not already part of the offensive player's group, it must eventually be moved. An expert player using this tactic can sometimes defer movement of their sentinel such that it becomes the final winning move. Note that just as in a "fork", a "block" can prevent multiple pieces from connecting.

9. **Tempo** – It's important not to get too far behind in the game. Although tempo is critical during all phases of the game, an advanced player can obtain a good sense of tempo during the end game by counting the number of moves required for each player to turn their largest group into a cluster.

Figure 15 illustrates a few of these concepts. White's attempt at forming a cluster is inhibited by the fact that it is bounded by the south and east edges of the board. Consequently, white has a limited number of liberties available. Furthermore, black has a sentinel which prevents white's group from becoming a cluster. White's only option is to relocate its group to another place on the board. However, doing so would result in white losing tempo, a serious disadvantage since black requires comparatively few remaining moves to win. This example also highlights the pitfall of increasing one's group size at the expense of sacrificing the liberties required to eventually form a cluster.

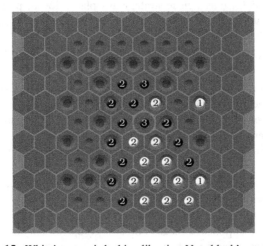

Fig. 15. White's group is lacking liberties. Note black's sentinel.

7 Experimental Game Designs

Included here is a sampling of a few experimental game designs considered, but not adopted since the goal was to converge on a single "best" version of Cluster. Playtesting revealed that although these game variations were both viable and playable, each had its shortcomings. However, if producing a "suite" of Cluster game variations had instead been the goal, then these games might serve as potential candidates.

1. **"Cluster-82"** (Fig. 16) – A large version of the game. It served as a starting point for paring down the size and number of pieces until the "optimal" configuration of the Cluster-58 version was finally settled upon.

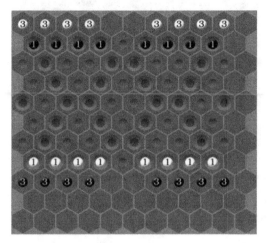

Fig. 16. Cluster-82.

2. **"Cluster-80"** (Fig. 17) – A large version of the game played on a hexagonal board. It was found that A ratio of 80 spaces to 22 pieces does not work well because the players often tend to avoid each other and the game becomes more of a race than a "thought provoking" positional game

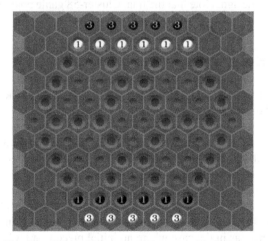

Fig. 17. Cluster-80 featuring a hexagonal hole Pattern.

3. **"Cluster-30"** (Fig. 18) – A "mini" version of the game. Due to the small number of holes, there is much less flexibility to the games.

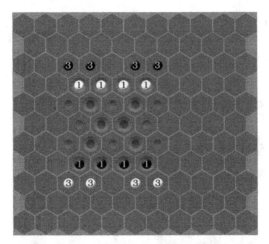

Fig. 18. Cluster-30, an experimental "mini" variation.

8   Piece to Space Ratio

An important discovery is that Cluster works much better and becomes more enjoyable when there is a higher level of interaction between the players' pieces. When there is a higher degree of scarcity of spaces at both hole depths, this results in increased competition between the players and they are required to do more strategic and tactical planning in order to form a cluster. Via extensive playtesting, a 2:1 ratio between game board spaces and total number of pieces appears to work best. The main design challenge was to discover a board topology that approximates this ratio while yielding a symmetrical and elegant aesthetic design. Note that the final Cluster-58 game board has a 2.23:1 ratio of spaces to pieces.

9   Design Heuristic Synopsis

As discussed, there were a number of heuristics that motivated the sequence of the Cluster designs. The most important ones are summarized below:

- **Board topology** – Refers to the overall shape of the board and how it affects game play. It includes effective distribution of the hole depths, consideration of the effect of gaps, and removal of infrequently used spaces.
- **Piece count and initial piece placement** – Refers to the number and initial placement of the pieces. Typically it has the largest effect during the opening phase of the game. However, it can also lead to longer range effects such as requiring all pieces to be moved at least once (as in the case where no initial piece is at level 2).
- **Piece to space ratio** - Affects the level of interaction of the pieces as well as the tempo of the game.
- **Rule simplicity** – The goal is to ensure that the rules are not unnecessarily complex.

- **Minimizing the potential for draws** – Ideally, most if not all games result in a clear winner. The rules, the board topology, and the piece count can have significant impact on this goal.
- **First player advantage** – The goal is to ensure that the second player can make a comparably strong reply to the first player's initial move.
- **Balance** – One player should not be able to easily "tip the scales" too easily such that the other player cannot recover.
- **Aesthetics & symmetry** – A subjective consideration which can often lead to aesthetically pleasing and balanced designs. It can positively affect both game play and player satisfaction.
- **Use of visual cues** – The goal is to reduce the player's mental burden on the game mechanics, thereby allowing greater focus on the game play itself.

Note that the above design heuristics are general enough that, in many cases, they may be applicable to games other than Cluster.

10   Game Design Wisdom

Each game design has its own key core elements that must be discovered and exploited in order to optimize the game play experience. Finding and refining these elements is arguably a blend of art and science. Aesthetics and intuition apply mainly to the "art" aspect, whereas applying one's knowledge base of game design heuristics, coupled with extensive playtesting, apply to the more formal "science" side. In retrospect, the Cluster design experience has revealed the following "wisdom" for approaching game design.

1. Don't quit or finalize your design too soon. In other words, your game may be good, but it might not be in its "preferred form".
2. Be open to making changes. Keep working until it cannot be improved any further. Be obsessed about your project and stay with it until it "feels" right.
3. Simplify the rules, and the format. Strive for elegance and avoid complexities, i.e. "less is more".
4. Be sure to thoroughly playtest your game. If there are any "flaws" either in the rules or in the topology of the game board, then determine and identify the problem and make changes to remove them. Don't "fall in love" with your first creation, as it most likely can be improved.
5. Try to holistically refine your game in order to find the optimum combination of rules, format, materials, sizes, shapes, colors, etc. until you feel that it cannot be improved any more. Clarity and aesthetics do matter here.

11   Conclusion

New game designs often arise by combining existing game concepts, and in the case of Cluster, may involve puzzle concepts as well. Recombination represents a powerful tool for game invention. Once the basis for the new game has been established, further refinement based on applying a variety of game design heuristics can be considered and then evaluated through playtesting.

In the case of Cluster, the variations are primarily based on the form of the shallow/deep hole configuration along with the initial setup of the pegs. These various setups can only be effectively evaluated via persistent experimentation and extensive playtesting.

The utilization of computer assisted game design software has so far provided two major benefits. Not only has it eliminated the need for physical prototypes, it has also accelerated the playtesting phase of proposed variations thereby promoting the discovery of new candidate variations.

Future work should explore this potential further. For example, through computer self-play that facilitates logging and replay of a series of games, various metrics of a specific game variation can be examined and assessed. For example, these metrics may include the degree of first player advantage and the average number of moves per game. Additionally, observing the automated games in a "replay" fashion may likely yield further strategic insights into effective game play.

As we have witnessed with Cluster, the improvement process may even span decades. As new insights are found, improved variations of the game are discovered constituting a "plateau" or "sub-optima" in the game's fitness landscape (e.g. an extreme example is the evolution of Chess rules over a period of centuries[7]). Finally, this design experience has revealed some general "wisdom", useful for approaching game design. The evolution of Cluster from Cluster-46 to Cluster-64, and ultimately to the current Cluster-58, offers an excellent example of applying these processes and principles.

Acknowledgements. Thanks to Cameron Browne for his helpful suggestions.

References

1. Sackson, S.: A Gamut of Games. Pantheon Books, New York (1982). [1st Pub. 1969, Random House, New York]. ISBN 0-394-71115-7
2. Hom, V., Marks, J.: Automatic design of balanced board games. In: Proceedings of the AAAI Conference on Artificial Intelligence and Interactive Digital Entertainment (AIIDE), pp. 25–30 (2007)
3. Browne, C., Maire, F.: Evolutionary game design. IEEE Trans. Comput. Intell. AI Games **2**(1), 1–16 (2010)
4. Shoptaugh, P.: Cluster game. U.S. Patent 3,834,708, filed February 23, 1973, and issued September 10, 1974 (1973)
5. Schmidt, G.: The axiom general purpose game playing system. IEEE Trans. Comput. Intell. AI Games **6**(4), 332–342 (2014)

---

[7] See http://www.chess.com/groups/forumview/history-of-chess-rules

Improving Human Players' T-Spin Skills in Tetris with Procedural Problem Generation

Taishi Oikawa, Chu-Hsuan Hsueh, and Kokolo Ikeda[✉]

School of Information Science, Japan Advanced Institute of Science and Technology, Nomi, Ishikawa, Japan
{taishi_o,hsuehch,kokolo}@jaist.ac.jp

Abstract. Researchers in the field of computer games interest in creating not only strong game-playing programs, but also programs that can entertain or teach human players. One of the branches is procedural content generation, aiming to generate game contents such as maps, stories, and puzzles automatically. In this paper, automatically generated puzzles are used to assist human players in improving the playing skills for the game of Tetris, a famous and popular tile-matching game. More specifically, a powerful technique called T-spin is hard for beginners to learn. To assist beginners in mastering the technique, automatically generated two-step to T-spin problems are given for them to solve. Experiments show that the overall ability for beginners to complete T-spin during play is improved after trained by the given problems. The result demonstrates the possibility of using automatically generated problems to assist human players in improving their playing skills.

Keywords: Procedural content generation · Puzzle · Tetris · Training system · Entertainment

1 Introduction

In recent years, artificial intelligence has made significant progress in computer board games as well as video games. Remarkably, computer programs can learn to surpass human levels without human knowledge of games. One example is the AlphaZero algorithm [17], which achieved superhuman levels of plays in three classical board games, chess, shogi, and Go. Another example for video games is that the programs based on deep reinforcement learning [8,10] obtained higher scores than professional human players in many of the Atari 2600 games.

Creating programs to entertain or teach human players is another popular research topic in the past decades. Hunicke and Chapman [4] proposed a probabilistic method to dynamically adjust the difficulty of a first-person shooter game. Ikeda and Viennot [6], and Sephton et al. [15] tried to create entertaining programs based on Monte-Carlo tree search (MCTS). Sephton et al. [15],

T. Cazenave et al. (Eds.): ACG 2019, LNCS 12516, pp. 41–52, 2020.
https://doi.org/10.1007/978-3-030-65883-0_4

Demediuk *et al.* [2], and Wu *et al.* [19] investigated strength and difficulty adjustment for MCTS-based programs. Ikeda *et al.* [5], Takahashi [18], and Oikawa and Ikeda [11] aimed to build training systems for the games of Go, Puyo-Puyo, and Tetris respectively.

This paper continues the work by Oikawa and Ikeda [11] to create a training system for Tetris, a famous tile-matching puzzle game. In Tetris, a technique called *T-spin* is powerful but hard to be learnt by beginners. As the first step to assist human players in mastering the technique, the previous work automatically generated *one-step to T-spin* problems. The generated problems were solved and rated by beginners in five-grade interestingness and difficulty. Generally, interestingness and difficulty were highly positively correlated.

In addition to generating problems, more emphasis in the paper is put on analyzing the effectiveness of training human players by the automatically generated problems. In the experiments, players are divided into three groups where one is trained by normal play and the other two by *two-step to T-spin problems* as well as normal play. After training, the win rates increase for all groups in a competition variant of Tetris. Especially, the two groups trained with T-spin problems can trigger T-spin more often. The results demonstrate the potential to build a training system for improving the playing skills of human players.

The rest of this paper is organized as follows. Subsect. 2 briefly introduces the game of Tetris, including the technique of T-spin, and then reviews some work related to procedural problem generation. Sections 3 and 4 present the two-stage experiments where the latter is the key experiments in this paper. In the first stage, prediction models for interestingness and difficulty of two-step problems are built. The second stage incorporates the interestingness predictor to generate problems to train human players. The experiments aim to verify how T-spin problems assist human players in improving their playing skills. Finally, Sect. 5 makes concluding remarks and discusses future research directions.

2   Background

This section introduces the game of Tetris in Subsection 2.1 and related work on procedural problem generation in Subsect. 2.2.

2.1   The Game of Tetris

Tetris is a kind of tile-matching puzzle video game presented in 1984 by Pajitnov[1]. The game has a widespread popularity and can be played on various platforms. There are many different variants of the game. This subsection briefly describes some basic elements of Tetris. The game is played on a rectangular field with twenty rows and ten columns. Seven kinds of *tetrominoes* are used, each of which is made up of four connected squares. The tetrominoes are named according to the similarity to Latin alphabet letters, I, J, L, O, T, S, and Z.

---

[1] https://en.wikipedia.org/wiki/Tetris.

During play, tetrominoes fall down by a random sequence. One tetromino falls down at a time, with the next one(s) shown as a hint. The player can move the given tetromino by one of the three directions left, right, and down, or rotate it by 90-degree. The tetromino stops falling when at least one square meets the bottom line or other occupied squares below, as shown by the bright gray squares in Figs. 1a and 1b in smaller fields. When all squares in a row are occupied, the row is cleared and the player obtains some bonus scores. An example of clearing a single row is depicted in Fig. 1c where the second row from the bottom is cleared. It is possible to clear at most four rows at a time, and higher scores are obtained when more rows are cleared. The squares above the cleared row(s) then fall down with the shape remained, as shown in Fig. 1d. A game ends when the occupied squares reach the top of the playing field and no tetromino can be further inserted. The goal of Tetris is to clear rows to obtain high scores. In some competition variants of Tetris, clearing rows on the player's own field generates *garbage* on the opponent field from the bottom, which may push the opponent closer to the end of games.

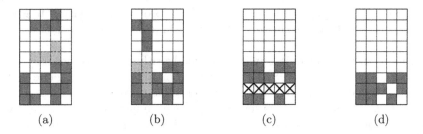

(a)   (b)   (c)   (d)

Fig. 1. An example of actions and row clearing in Tetris: (a) the current state, (b) manipulating the given L-tetromino, (c) clearing a row, and (d) the state after the row cleared.

A technique called *T-spin* is to insert a T-tetromino into a tight space by rotation immediately after the T-tetromino stops falling. An example of T-spin is shown in Fig. 2. For the state in Fig. 2a, if the T-tetromino falls down directly, a square in the second row from the bottom remains empty, as shown in Fig. 2b. At the moment, the T-tetromino can be rotated clockwise by 90-degree, and then the state becomes the one in Fig. 2c. In this example, two rows are cleared after T-spin, which is called *T-spin Double*. It is possible to clear at most three rows with T-spin, which is called *T-spin Triple*. Clearing rows with T-spin obtains extra bonuses; moreover, in competition variants of Tetris, more garbage is generated on the opponent field.

2.2 Related Work on Procedural Problem Generation

Procedural Content Generation (PCG) [16] is a research branch aiming to automatically create *game contents* by computer programs with limited or indirect human input. The term game content covers a large variety of components and

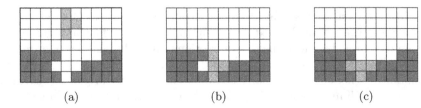

Fig. 2. An example of T-spin Double: (a) the current state, (b) falling down the T-tetromino, and (c) rotating the stopped T-tetromino to finish T-spin Double.

may refer to maps, levels, stories, and puzzle problems. This paper focuses on generating puzzle problems [1] and reviews some related work as follows.

Hirose *et al.* [3] proposed to compose tsume-shogi problems (mating problems in shogi) by *reverse method*, which started from board states with checkmate and then searched reversely for n moves. Two of the generated problems were introduced in a tsume-shogi magazine and received favourable feedback. For chess mating problems, Schlosser [14] automatically composed based on endgame tables built up by retrograde analysis. Iqbal [7] improved the efficiency to generate mate-in-3 problems with incorporating probabilities of piece locations derived from different databases.

Mantere and Koljonen [9] applied genetic algorithm (GA) to not only solve but also generate and rate Sudoku puzzles. For another puzzle game called Shinro, Oranchak [12] showed that GA was effective to construct puzzles. Ortiz-García *et al.* [13] proposed to automatically generate picture-logic puzzles (also known as nonograms) from RGB color images. Takahashi [18] compared random method and reverse method on generating problems for Puyo-Puyo, where the later was shown to be more efficient.

Oikawa and Ikeda [11] proposed to generate n-step to T-spin problems by reverse method. The approach is briefly summarized as follows. For the simplicity of discussion, T-spin in the rest of this paper refers to T-spin Double if not specified. First, basic patterns with a complete T-shape was created, as the part highlighted in Fig. 3a. To enrich the diversity of the generated T-spin patterns, noises were introduced. An example of the resulted T-spin pattern is depicted in Fig. 3a. Reverse method then generated n-step to T-spin problems by removing n tetrominoes from the playing field one at a time. Figures 3b and 3c show two examples of one-step to T-spin problems with removing the highlighted S-tetromino and J-tetromino respectively. A two-step to T-spin problem with S- and J-tetrominoes is shown in Fig. 3d. In their experiments, one-step to T-spin problems were generated and analyzed. They showed that interestingness and difficulty of the problems had a highly positive correlation for beginners.

3   Experiments on Interestingness and Difficulty

As a preliminary study, a total of sixteen players (males between 22 and 27 years) participated in the experiments. All of them were interested in playing games,

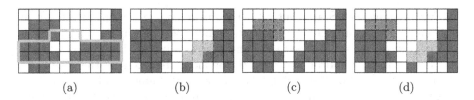

(a)   (b)   (c)   (d)

Fig. 3. An example of the reverse method for T-spin problems: (a) a T-spin pattern, (b) a one-step problem with S-tetromino, (c) a one-step problem with J-tetromino, and (d) a two-step problem with S- and J-tetrominoes.

among which one was an expert of Tetris and the rest were beginners. To verify how two-step to T-spin problems assisted in improving the playing skills of beginners, the experiments were designed with two parts. In the first part, as described in more details in this section, predictors on interestingness and difficulty of two-step problems were built by data collected from the beginners. In the second part, beginners were divided into three groups, where one was trained without T-spin problems, one with randomly generated problems, and the other with interesting problems. The details are included in Sect. 4.

In this section, the settings of experiments on interestingness and difficulty, the features extracted for two-step to T-spin problems, and the results are presented in Subsects. 3.1, 3.2, and 3.3 respectively.

3.1   Experiment Settings

First, the T-spin technique and its importance to Tetris were explained to the beginners. They then solved 50 two-step problems with the tool shown in Fig. 4a. All players solved the same set of problems but in random order.

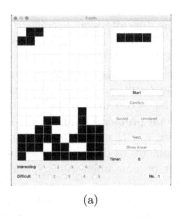

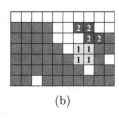

(b)

(a)

Fig. 4. (a) Tool for the experiments and (b) an example for the new features in two-step problems.

After pressing the "Start" button, 50 problems were given one at a time. At the beginning of a problem, the given tetromino was put in the playing field, with the next one shown on the right-hand side, as illustrated in Fig. 4a. Players could drag, rotate, and drop the given tetromino. After pressing the "Confirm" button, the location of the first tetromino was determined and could not be changed. The second tetromino was then put into the playing field. After determining the location of the second tetromino, players rated the interestingness and the difficulty in five-grade evaluation. Scores 1 to 5 represented most uninteresting/easiest to most interesting/difficult. Solutions could also be obtained by pressing the "Show Answer" button. Players then pressed the "Next" button to solve the next problem until all problems were finished. The reason to enable players to obtain the solutions was to prevent the players from giving up due to frustration and rating the problems arbitrarily.

3.2   Features of Two-Step to T-Spin Problems

Oikawa and Ikeda [11] have proposed some features for one-step problems. Two-step problems are supposed to be more difficult than one-step problems. Some problems may require the cooperation between the two given tetrominoes, as shown in Fig. 4b. In order to better predict the interestingness and difficulty, four additional features are designed in this paper: (1) the number of edges connected between the two given tetrominoes, (2) the number of squares of the second tetromino that are located above the first one, (3) the number of squares of the two given tetrominoes that are located in the two rows to be cleared by T-spin, and (4) the number of squares of the two given tetrominoes that are located below the T-shape. The values of the four features for the example in Fig. 4b are 1, 3, 4, and 2 respectively.

3.3   Results of Predictors

For each problem, the interestingness and the difficulty were averaged from the fifteen beginners. The results are plotted in Fig. 5a. The correlation coefficient between interestingness and difficulty was 0.86, which showed a highly positive correlation. When looking into the ratings player by player, different preference can be found. This paper comments on the results of two players, as shown in Figs. 5b and 5c with the data scattered for better display. For player A, no problems were rated as difficulty 5, and difficult problems tended to be interesting. Player B tended to rate difficult problems as either very interesting or very uninteresting. The player also rated easy problems as different interestingness.

It is expected that problems too difficult become uninteresting, and Player B did rate some difficult problems as very uninteresting. However, the overall interestingness and difficulty for two-step problems still had a high correlation. One possible explanation is that two-step problems were not too difficult for most of the beginners. The decrease of interestingness may occur for problems with more steps which are supposed to be even more difficult.

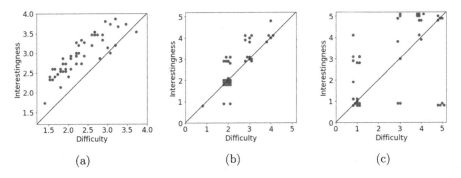

Fig. 5. Relations between interestingness and difficulty for (a) fifteen beginners averaged, (b) player A, and (c) player B.

To automatically generate interesting problems to train human players, a model to rate the interestingness of problems is required. In the experiments, supervised learning was applied to learn the interestingness. Ratings averaged from fifteen beginners for a total of 50 problems were used as the training data, with the features described in Subsect. 3.2. The predictor was built based on the LightGBM framework[2] with 10-fold cross-validation. The results are plotted in Fig. 6a, and the symmetric mean absolute percentage error (SMAPE) and mean absolute error (MAE) were 8.74% and 0.27 respectively. The model was able to predict the interestingness of two-step problems well. The results of the difficulty predictor are shown in Fig. 6b, with a little higher SMAPE of 14.50% and MAE of 0.34. The model performed better for easier problems.

Figure 7a shows a problem which was predicted to be interesting by the interestingness predictor. Players may feel that the center part is a little bit empty and doubt whether it is possible to complete T-spin. The two L-tetrominoes, no matter the order, both locate beside the T-spin pattern. Once the solution is found, players may feel more satisfied and interesting.

One additional experiment was to ask the expert to solve and rate 147 two-step to T-spin problems. The results in Fig. 6c shows that it was also possible to create interestingness predictors for individuals even experts, though the SMAPE of 26.17% and MAE of 0.84 were higher. One possible reason is that the granularity of the ratings for a single player is larger than the averaged values. Still, the model can be used to distinguish problems with interestingness 1 by predicted values lower than 2 and interestingness 4 and 5 by predicted values higher than 4.7. A problem considered interesting by the predictor is shown in Fig. 7b. At the first glance, the expert thought that T-spin could be completed. However, the way that the expert considered resulted in a T-spin Single, which was thought as a trap and made the expert feel interesting. The experiment demonstrated the potential to personalize training systems.

---

[2] https://lightgbm.readthedocs.io/en/latest/.

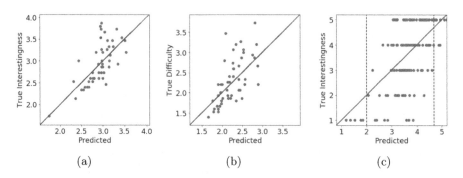

Fig. 6. Results of predictors on (a) interestingness and (b) difficulty for beginners, and (c) interestingness for an expert.

Fig. 7. Interesting problems for predictors on (a) beginners and (b) an expert.

4   Experiments on Improving Human Players' Skills

This section presents the key experiments in this paper to demonstrate how two-step to T-spin problems assist human players in improving their T-spin skills in Tetris. Subsection 4.1 describes the experiment settings. The results and some discussions are then included in Subsect. 4.2.

4.1   Experiment Settings

The flow chart of the experiments is shown in Fig. 8. At the beginning, the strength of the players was measured by two approaches. First, they played best-of-three matches against three built-in CPU agents (Arle, Ess, and Zed) in Puyo-Puyo Tetris on the Steam platform[3]. Puyo-Puyo Tetris is a competition variant of Tetris, and the three CPU agents have different strength. Each player played at least six and at most nine games. The numbers of wins and losses, the time, the number of T-spin, and the whole progress of the games were recorded. The average time to finish a best-of-three match was 448 s with the standard deviation of 231 s. The actual time varied a lot from players, with the shortest and longest time of 161 s and 1,354 s respectively.

In addition, the players solved fifteen two-step to T-spin problems selected by an expert. Each of the three levels, easy, middle, and difficult from the expert's

---

[3] https://store.steampowered.com/app/546050/Puyo_PuyoTetris.

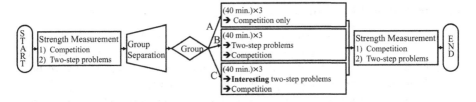

Fig. 8. Flow chart of the experiments on training human players.

view, contained five problems. The tool for solving problems was similar to the one shown in Fig. 4a, except that the time limit for each problem was set to 30 s and the players could not obtain the solutions.

To investigate whether T-spin problems assisted players in improving playing skills, three different kinds of training menus were designed, one without T-spin problems while the other two with T-spin problems. Fifteen players were evenly divided into three groups such that the strength of the groups was similar. By strength of a group, it was the sum of the number of wins against the CPU agents and the number of correctly solved problems. The numbers are listed in Table 1, where "#W", "#G", "F_{TS}", and "#Sol" represent the number of wins, the number of played games, the frequency of T-spin per minute during play, and the number of correctly solved problems respectively. The three menus were given to the groups randomly. One set of training lasted for 40 min. All groups underwent the given sets of training three times. Namely, all the players were trained for 120 min.

Table 1. Results for groups A, B, and C before and after training.

	A (Competition Only)				B (Two-Step Problems)				C (Interesting Problems)			
	#W	#G	F_{TS}	#Sol	#W	#G	F_{TS}	#Sol	#W	#G	F_{TS}	#Sol
Before	7	35	0.21	59	4	33	0.24	62	5	32	0.18	61
After	13	33	0.13	60	11	36	0.75	67	8	33	0.79	61

The players in group A played against all three CPU agents Arle, Ess, and Zed during the whole 40 min in one set of training. For groups B and C, players were trained by 75 two-step to T-spin problems first. A similar tool to Fig. 4a was used except that the time limit for each problem was set to 20 s and the solutions were always displayed after the players solved the problems. Group B was provided with problems randomly generated, while group C with interesting problems rated by predictor for beginners. Training by T-spin problems took about 20 min. In the rest of the time in 40 min, the players in both groups played against CPU agents as group A. After three sets of training were finished, the strength of the players was measured again, with the order of the fifteen problems changed.

4.2   Results of Skill Improvement

The results for the three groups after training are listed in Table 1. The win rates of all the groups increased after training, which were 20.0±13.3% to 39.4±16.7%, 12.1±11.1% to 30.6±15.0%, and 15.6±12.6% to 24.2±14.6% respectively. Although the result of group C was not statistically significant, the experiments still showed that the strength of the players with regard to the competition variant could be improved after training. The correctness rates for T-spin problems of 78.7±9.3% to 80.0±9.1%, 82.7±8.6% to 89.3±7.0%, and 81.3±8.8% to 81.3±8.8% were not improved significantly, even for groups B and C.

However, when focusing on the frequency of T-spin, the growth rates for groups B and C were 212% and 334% respectively, while group A was −39%. The frequencies of T-spin per minute for the fifteen players are plotted in Fig. 9. Generally, players in groups B and C improved their T-spin skills. Especially, for those who completed less T-spin before training, the growth was even clearer. Interestingly, a player in group A tended not to perform T-spin after training.

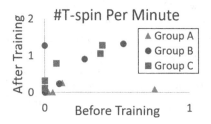

Fig. 9. Frequency of T-spin per minute before and after training.

To investigate the reason why players with higher T-spin skills did not win more games (groups B and C vs. group A), the recorded progress of the games was reviewed by an expert. It was found that the speeds to arrange the tetrominoes were reduced while the players attempted to create T-spin patterns. Sometimes, they also made mistakes on the arrangement of tetrominoes. As a result, the merits obtained from T-spin were almost eliminated by the speed reduction and the mistakes. If the players were trained for a longer time to be familiar with T-spin patterns, their strength may improve more significantly. Overall, the experiments suggested that the strength of players could be improved after training. Especially, players trained with T-spin problems improved the strength with aware of the technique.

5   Conclusions and Future Work

In this paper, two-step to T-spin problems are generated automatically to train beginners. First, models trained by ratings collected from beginners can predict the interestingness of two-step problems well. In addition, it is also possible to train models to predict interestingness of problems for individual players. This demonstrates the potential to personalize training systems of games.

Moreover, experiments on training players with three different menus are conducted. Overall, the strength of the players is improved. Especially, players trained with T-spin problems indeed complete T-spin more frequently, though they do not improve win rates more than the players trained without T-spin problems. It is supposed that the difference will be bigger if the training time became longer for the players to master their T-spin skills.

The followings discuss several promising research directions. With the success in predicting the interestingness of one-step and two-step to T-spin problems. the next step to the training system is to try three-step or four-step problems, which are even closer to real game-play. It is also interesting to find the point that difficult problems become uninteresting for beginners. In addition, some more techniques such as clearing four rows and T-spin Triple are also important to Tetris. Similar approaches can be applied to build training systems for these techniques. It is also possible to mix different techniques in a training system.

Acknowledgments. This research is financially supported by Japan Society for the Promotion of Science (JSPS) under contract numbers 18H03347 and 17K00506.

References

1. De Kegel, B., Haahr, M.: Procedural puzzle generation: a survey. IEEE Trans. Games **12**(1), 21–40 (2020). https://doi.org/10.1109/TG.2019.2917792
2. Demediuk, S., Tamassia, M., Raffe, W.L., Zambetta, F., Li, X., Mueller, F.: Monte Carlo tree search based algorithms for dynamic difficulty adjustment. In: 2017 IEEE Conference on Computational Intelligence and Games (CIG 2017), pp. 53–59. IEEE (2017). https://doi.org/10.1109/CIG.2017.8080415
3. Hirose, M., Ito, T., Matsubara, H.: Automatic composition of Tsume-shogi by reverse method. J. Jpn. Soc. Artif. Int. **13**(3), 452–460 (1998)
4. Hunicke, R., Chapman, V.: AI for dynamic difficulty adjustment in games. In: AAAI-04 Workshop on Challenges in Game Artificial Intelligence, pp. 91–96. AAAI Press (2004)
5. Ikeda, K., Shishido, T., Viennot, S.: Machine-learning of shape names for the game of Go. In: Plaat, A., van den Herik, J., Kosters, W. (eds.) ACG 2015. LNCS, vol. 9525, pp. 247–259. Springer, Cham (2015). https://doi.org/10.1007/978-3-319-27992-3_22
6. Ikeda, K., Viennot, S.: Production of various strategies and position control for Monte-Carlo Go - entertaining human players. In: 2013 IEEE Conference on Computational Intelligence in Games (CIG 2013), pp. 145–152. IEEE (2013). https://doi.org/10.1109/CIG.2013.6633625
7. Iqbal, A.: Increasing efficiency and quality in the automatic composition of three-move mate problems. In: Anacleto, J.C., Fels, S., Graham, N., Kapralos, B., Saif El-Nasr, M., Stanley, K. (eds.) ICEC 2011. LNCS, vol. 6972, pp. 186–197. Springer, Heidelberg (2011). https://doi.org/10.1007/978-3-642-24500-8_20
8. Kapturowski, S., Ostrovski, G., Quan, J., Munos, R., Dabney, W.: Recurrent experience replay in distributed reinforcement learning. In: The Seventh International Conference on Learning Representations (ICLR 2019) (2019)

9. Mantere, T., Koljonen, J.: Solving, rating and generating Sudoku puzzles with GA. In: 2007 IEEE Congress on Evolutionary Computation (CEC 2007), pp. 1382–1389. IEEE (2007). https://doi.org/10.1109/CEC.2007.4424632

10. Mnih, V., et al.: Human-level control through deep reinforcement learning. Nature **518**(7540), 529–533 (2015). https://doi.org/10.1038/nature14236

11. Oikawa, T., Ikeda, K.: Procedural problem generation of Tetris for improving T-spin skill. In: The 23rd Game Programming Workshop (GPW-18), pp. 175–182 (2018)

12. Oranchak, D.: Evolutionary algorithm for generation of entertaining Shinro logic puzzles. In: Di Chio, C., et al. (eds.) EvoApplications 2010. LNCS, vol. 6024, pp. 181–190. Springer, Heidelberg (2010). https://doi.org/10.1007/978-3-642-12239-2_19

13. Ortiz-García, E.G., Salcedo-Sanz, S., Leiva-Murillo, J.M., Pèrez-Bellido, A.M., Portilla-Figueras, J.A.: Automated generation and visualization of picture-logic puzzles. Comput. Graph. **31**(5), 750–760 (2007). https://doi.org/10.1016/j.cag.2007.08.006

14. Schlosser, M.: Computers and chess problem composition. ICCA Journal **11**(4), 151–155 (1988). https://doi.org/10.3233/ICG-1988-11404

15. Sephton, N., Cowling, P.I., Slaven, N.H.: An experimental study of action selection mechanisms to create an entertaining opponent. In: 2015 IEEE Conference on Computational Intelligence and Games (CIG 2015), pp. 122–129. IEEE (2015). https://doi.org/10.1109/CIG.2015.7317939

16. Shaker, N., Togelius, J., Nelson, M.J.: Procedural Content Generation in Games: A Textbook and an Overview of Current Research. Springer (2016)

17. Silver, D., et al.: A general reinforcement learning algorithm that masters chess, shogi, and Go through self-play. Science **352**(6419), 1140–1144 (2018). https://doi.org/10.1126/science.aar6404

18. Takahashi, R.: Mating problem generation of Puyo-Puyo for training. Master thesis, Japan Advanced Institute of Science and Technology, Nomi, Ishikawa, Japan (2018)

19. Wu, I.C., Wu, T.R., Liu, A.J., Guei, H., Wei, T.h.: On strength adjustment for MCTS-based programs. In: The 33rd AAAI Conference on Artificial Intelligence (AAAI-19). AAAI Press (2019)

A Further Investigation of Neural Network Players for Game 2048

Kiminori Matsuzaki[✉]

School of Information, Kochi University of Technology, Kami, Kochi 782–8502, Japan
matsuzaki.kiminori@kochi-tech.ac.jp

Abstract. Game 2048 is a stochastic single-player game. Development of strong computer players for Game 2048 has been based on N-tuple networks trained by reinforcement learning. Some computer players were developed with neural networks, but their performance was poor. In our previous work, we showed that we can develop better policy-network players by supervised learning. In this study, we further investigate neural-network players for Game 2048 in two aspects. Firstly, we focus on the component (i.e., layers) of the networks and achieve better performance in a similar setting. Secondly, we change input and/or output of the networks for better performance. The best neural-network player achieved average score 215 803 without search techniques, which is comparable to N-tuple-network players.

Keywords: Game 2048 · Neural network · Supervised learning

1 Introduction

Neural networks (NN) are now widely used in development of computer game players. Among these, deep convolutional neural networks have been studied actively in recent years and played an important role in the development of master-level computer players, for example, for Go (AlphaGo Zero [11]), Chess (Giraffe [7], DeepChess [2] and AlphaZero [10]) and Shogi (AlphaZero [10]).

The target of this study is Game "2048" [1], a stochastic single-player game. Game 2048 is a slide-and-merge game and its "easy to learn but hard to master" characteristics have attracted quite a few people. According to its author, during the first three weeks after its release, people spent a total time of over 3000 years on playing the game.

Several computer players have been developed for Game 2048. Among them, the most successful approach is to use N-tuple networks as evaluation functions and apply a reinforcement learning method to adjust the weights in the networks. This approach was first introduced to Game 2048 by Szubert and Jaśkowski [12], and several studies were then based on it. The state-of-the-art computer player developed by Jaśkowski [5] combined several techniques and achieved an average score of 609 104 within a time limit of 1 s per move.

© Springer Nature Switzerland AG 2020
T. Cazenave et al. (Eds.): ACG 2019, LNCS 12516, pp. 53–65, 2020.
https://doi.org/10.1007/978-3-030-65883-0_5

NN-based computer players, however, have not achieved a success yet for Game 2048. The first work by Guei et al. [4] proposed a player with two convolution layers followed by two full-connect layers, but the average score was about 11 400. The player by tjwei [13] used two convolution layers with a large number of weights, and achieved an average score 85 351 after supervised learning. Though there exist other implementations of NN-based players [9,14,16], the scores of these players were not so good or were not reported. In our previous work [6], we tried to improve the performance of NN-based players by increasing the number of convolution layers and supervised learning. We designed networks with 2–9 convolution layers with 2×2 filters and applied supervised learning with playlogs from existing strong players [8]. As the result, we achieved the best average score 93 830 with a player with five convolution layers.

In this paper, we further investigate the NN-based players in two aspects. Firstly, we focus on the component of the networks. We compare players with full-connect layers (multi-layer perceptrons), convolution layers with 2×2 filters, and convolution layers with 1×2 filters. We also compare the performance of these players by using two sets of playlogs from two different players. Secondly, we explore the input and/or output of networks. We designed four networks by changing the input and output of the networks. The best player obtained in the paper achieved an average score 215 803 without search techniques.

Contributions in this paper are summarized as follows.

- We designed three networks by changing the components (layers). The experiment results showed that the network with convolution layers with 1×2 filters performed the best. (Sect. 4)
- We used two sets of playlogs from two different players as the training data. We confirmed by experiments that the performance of players could differ significantly even though the average and maximum scores of the playlogs were almost the same. (Sect. 4)
- We designed four networks by changing input and/or output. The experiment results showed that a straightforward implementation of value network does not work well. The network named Policy AS performed the best, and the best average score was 215 803, which is comparable to greedy plays of N-tuple-network players. (Sect. 5)

The rest of the paper is organized as follows. Section 2 briefly introduces the rule of Game 2048. Section 3 reviews related work in terms of existing neural-network player for Game 2048. Section 4 designs neural networks with different components where we find that convolution layers with 1×2 filters perform the best. Section 5 designs neural networks with different input/output where we find that policy networks can be improved by extending the input. We conclude the paper in Sect. 6.

2   Game 2048

Game 2048 is played on a 4×4 grid. The objective of the original Game 2048 is to reach a 2048 tile by moving and merging the tiles on the board according to

(a) An initial state. Two tiles are placed randomly.

(b) After the first move: *up*. A new 2-tile appears at the lower-left corner.

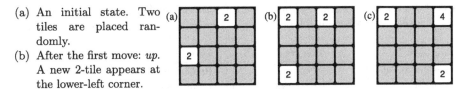

(c) After the second move: *right*. Two 2-tiles are merged to a 4-tile, and score 4 is given. A new tile appears at the upper-left corner.

Fig. 1. Process of game 2048

the rules below. In an initial state (Fig. 1), two tiles are placed randomly with numbers 2 ($p_2 = 0.9$) or 4 ($p_4 = 0.1$). The player selects a direction (either up, right, down, or left), and then all the tiles will move in the selected direction. When two tiles of the same number collide, they create a tile with the sum value and the player gets the sum as the score. Here, the merges occur from the far side and newly created tiles do not merge again on the same move: move to the right from 222␣, ␣422 and 2222 results in ␣␣24, ␣␣44, and ␣␣44, respectively. Note that the player cannot select a direction in which no tiles move nor merge. After each move, a new tile appears randomly at an empty cell with number 2 ($p_2 = 0.9$) or 4 ($p_4 = 0.1$). If the player cannot move the tiles, the game ends.

When we reach the first 1024-tile, the score is about 10 000. Similarly, the score is about 21 000 for a 2048-tile, 46 000 for a 4096-tile, 100 000 for an 8192-tile, 220 000 for a 16384-tile, and 480 000 for a 32768-tile.

3   Related Work: Neural-Network Players for Game 2048

Several computer players have been developed for Game 2048. The most widely used and successful approach is based on N-tuple networks trained by reinforcement learning methods [5,8,12,15]. The state-of-the-art player by Jaśkowski [5] was based on a redundant N-tuple network that was adjusted by the temporal coherence learning with some other techniques. It achieved an average score 324 710 with the greedy (1-ply search) play and 609 104 with the expectimax search within a time limit of 1 s per move.

Behind the success of N-tuple networks, neural networks have not been actively studied or utilized for the development of Game 2048 players. As far as the authors know, the work by Guei et al. [4] was the first study on the subject. Some open-source programs have been developed, for instance, with multi-layer perceptrons [16], with convolutional neural networks [13,14] and with recurrent neural networks [9], but the performance of these players was not very good or not analyzed well (at least from the documents provided). Table 1 summarizes existing neural-network players in terms of the number of weights and the average score of greedy (1-ply search) play.

In our previous work [6], we tried to improve the performance of neural-network players by increasing the number of convolution layers. We designed

Table 1. Summary of neural-network players in terms of number of weights and average score of greedy (1-ply search) play.

Authors	Network	Learning	Weights	Score
Guei et al. [4]	2 conv (2 × 2), 2 FC	Reinforcement	N/A	11 400*
Guei et al. [4]	3 conv (3 × 3), 2 FC	Reinforcement	N/A	5 300*
tjwei [13]	2 conv (2 × 1), 1 FC	Supervised	16.95×10^6	85 351
tjwei [13]	2 conv (2 × 1), 1 FC	Reinforcement	16.95×10^6	33 000*
Virdee [14]	2 conv (2 × 1), 2 FC	Reinforcement	1.98×10^6	16 000*
Kondo and Matsuzaki [6]	2 conv (2 × 2), 1 FC	Supervised	0.82×10^6	25 669
Kondo and Matsuzaki [6]	5 conv (2 × 2), 1 FC	Supervised	0.82×10^6	93 830
This work	3 conv (2 × 1), 2 FC	Supervised	3.69×10^6	215 802

conv = convolution layers, FC = full-connect layers. Values marked by * were read from figures.

networks with 2–9 convolution layers with 2 × 2 filters and applied supervised learning with playlogs from existing strong players [8]. As the result, we achieved better results than existing neural-network players and the best player with five convolution layers achieved an average score 93 830.

4   Experiment 1: Changing Components

Network Structure. The first experiment is about the components (layers) of neural networks. We design a network with full-connect layers only (i.e., multi-layer perceptrons) and two networks with convolution layers as shown in Fig. 2. In this experiment, we borrow the basic design of the networks from our previous work [6]. The input board is a binary-encoded 16-channel image such that the first channel represents positions of empty cells, the second does 2-tiles, the third does 4-tiles, and so on. The output consists of 4 values each of which represents probability of selecting the corresponding direction (i.e., policy network).

The most basic components in neural networks are full-connect layers and convolution layers. Convolution layers are often used in many applications because we can reduce the number of parameters by capturing local features. However, since the board size of Game 2048 is just 4 × 4, the effect of parameter reduction would be small. In contrast, use of full-connect layers from the first layer would have a benefit of directly capturing some global features. Therefore, we designed network MLP that consists of five full-connect layers (Fig. 2 (a)). In the network MLP, we increased the number of filters as 512, 1024, 2048, in order, and then narrowed down to 256 at the fourth layer. We conducted preliminary experiments by changing the number of filters and encountered failures of training when a large number of filters (e.g., 1024) are used at the fourth layer. The last layer is a full-connect layer with a softmax function. We did not use dropouts in our experiments.

The second network CNN22 has two convolution layers with 2 × 2 filters. 2 × 2 filters were used in the work by Guei et al. [4] and in our previous work [6]. In our design, the network has two convolution layers in which 2 × 2 filters are applied

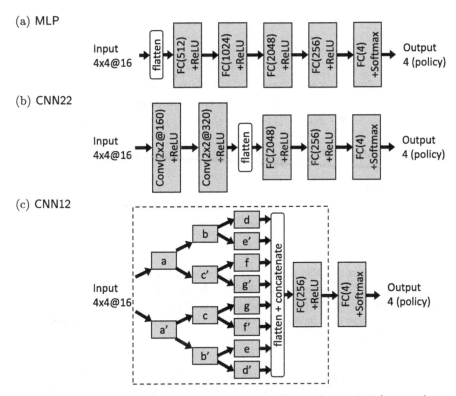

Every component in the first three layers has 256 filters of size 1×2 (or 2×1). Component x' is the transposition of x and their weights are shared.

Fig. 2. Structure of networks used in first experiment

without padding (yielding 2×2 outputs). Considering the number of parameters and GPU memory usage, we set the number of filters for the convolution layers to be 160 and 320, in order. After the convolution layers, three full-connect layers follow in the same manner as the network MLP.

The third network CNN12 has three convolution layers with 1×2 (and 2×1) filters. These 1×2 and 2×1 filters were also used by tjwei [13]. The differences are (1) convolution layers are in three plies, (2) only the outputs of the third layer are concatenated (tjwei's network concatenated all the outputs of two layers), and (3) we share weights between filters by considering duality to reduce the number of parameters. For instance in Fig. 2, filters in a' (of size 2×1) are transposes of those in a (of size 1×2). In the second and third layers, similarly, the filters with the same label share weights through transposition. Each component in the first three layers has 256 filters, and thus we have 256, 512, and 1024 filters in total, in order. After the three convolution layers, we flatten and concatenate the outputs and apply two full-connect layers in the same way as the network MLP.

Table 2. Size and cost of neural networks used in the first experiment.

Name	# parameters	Learning		Playing	
		Time	GPU Memory	Time	GPU Memory
MLP	3 281 668	57 min	377 MiB	0.57 ms	377 MiB
CNN22	3 364 580	66 min	6 135 MiB	1.24 ms	887 MiB
CNN12	3 681 284	117 min	3 181 MiB	1.95 ms	765 MiB

Table 3. Players used for generating training dataset.

Name	Description	Ave. score	Max. score
acg17 [8]	4 × 6-tuples, 8 stages, 3-ply expectimax	461 062	838 332
nneo [17]	hand-made features, variable-depth expectimax	435 448	840 320

Table 2 summarizes the number of parameters, execution time and GPU memory usage for the learning and playing phases[1]. As we can see in the table, though the number of parameters was controlled to be almost the same, the GPU memory usage differed largely: CNN22 consumed 16 times as large GPU memory as MLP did.

Supervised Learning. As in our previous work [6], we apply supervised learning to train neural networks. We collected a large number of playlogs from two strong computer players and utilized the players' moves as the correct labels. The training was performed to minimize the cross entropy.

To see the impact of training datasets, we prepared two datasets. The first one named acg17 is a set of playlogs from players developed in our previous work [8]. The players utilized N-tuple networks as the evaluation functions: the networks consisted of four 6-tuples and the game was split into eight stages based on the maximum number of tiles. The weights of the networks were adjusted by backward temporal coherent learning with restart strategy. The player selected moves by the 3-ply expectimax search. The second one named nneo is playlogs from an open-source player [17]. It utilized hand-made features and variable-depth expectimax search.

We collected for each player more than 6×10^8 actions and then chose 5×10^8 actions for the training and different 10 000 actions for the evaluation. The training data were shuffled before fed to the training. The average and maximum scores of these two players are given in Table 3. The performance of these two players seems to be almost the same.

Playing Method. Since the board of Game 2048 is symmetric, we can fed symmetric boards to a network to make robust decisions. In our playing method,

---

[1] All the experiments were conducted on a single PC with an Intel Core i3-8100 CPU, 16 GB Memory, and a GeForce GTX 1080 Ti GPU (GPU Memory 11 GB).

Table 4. Evaluation results of the first experiment for dataset `acg17`.

Name		Loss	Agreement [%]	Score
MLP	1×10^8 actions	0.571 ± 0.019	73.31 ± 0.18	$33\,543 \pm 5\,209$
	2×10^8 actions	0.544 ± 0.018	74.64 ± 0.23	$37\,159 \pm 6\,249$
CNN22	1×10^8 actions	0.554 ± 0.017	74.07 ± 0.25	$63\,969 \pm 5\,163$
	2×10^8 actions	0.526 ± 0.017	75.29 ± 0.26	$74\,867 \pm 3\,746$
CNN12	1×10^8 actions	0.535 ± 0.018	75.63 ± 0.63	$77\,074 \pm 6\,351$
	2×10^8 actions	0.513 ± 0.027	75.92 ± 0.63	$91\,728 \pm 5\,609$

Table 5. Evaluation results of the first experiment for dataset `nneo`.

Name		Loss	Agreement [%]	Score
MLP	1×10^8 actions	0.464 ± 0.009	79.34 ± 0.39	$6\,911 \pm\ \ 320$
	2×10^8 actions	0.433 ± 0.013	81.39 ± 0.48	$8\,201 \pm\ \ 829$
CNN22	1×10^8 actions	0.447 ± 0.017	80.88 ± 0.25	$14\,935 \pm\ \ 750$
	2×10^8 actions	0.383 ± 0.020	82.70 ± 0.18	$18\,258 \pm 1\,587$
CNN12	1×10^8 actions	0.423 ± 0.018	81.57 ± 0.33	$25\,694 \pm 2\,311$
	2×10^8 actions	0.387 ± 0.027	83.20 ± 0.33	$30\,810 \pm 2\,429$

we fed eight symmetric boards and pick up the direction with the majority vote. If two or more directions have ties, we select the direction based on the sum of probabilities. Refer to [6] for the details of the playing method.

Feeding symmetric boards at the playing improved the performance largely. In contrast, we did not observe a significant difference when we fed symmetric boards during the training.

Evaluation. We conducted five runs (of training and evaluation) for each network structure and dataset. For the training, we selected 2×10^8 actions from shuffled playlogs and fed them to the training with batches of size $1\,000$. We observed the values of loss functions as a metric of training progress. For the evaluation, we computed the direction for each evaluation board and calculated the agreement ratio to the dataset player. We also played 100 games and calculated the average score.

Table 4 shows the experiment results for the dataset `acg17` after learning 1×10^8 and 2×10^8 actions, and Table 5 for the dataset `nneo`. Note that the standard deviations over five runs are denoted after the $\pm$ sign. Figures 3 and 4 plot the agreement ratio and the average score of test plays, respectively.

Comparing the three networks, CNN12 with the smallest filters performed the best and MLP performed the worst. The difference of the loss values was small: 0.018 between MLP and CNN22; 0.013 between CNN22 and CNN12. The agreement ratios differ in some degree (about 1–2%). However, the scores of test

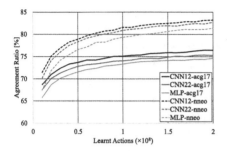

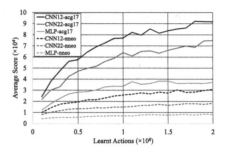

Fig. 3. Agreement ratio of the first experiment.

Fig. 4. Average score of the first experiment.

plays differ considerably: by a factor of 1.2 between CNN12 and CNN22; by a factor of 2.3 between CNN12 and MLP.

Another interesting finding is that though the two datasets seemed similar in terms of the average and maximum scores, the training with them resulted in very different players. In particular, the loss values and the agreement ratios for the dataset nneo were better than those for the dataset acg17, but the player trained with the dataset nneo performed much worse. We guess that the difference came from the difference in play styles. Detailed analysis remains as our future work.

5   Experiment 2: Changing Input/Output

Since we have obtained a good structure for the policy network in the first experiment, we now extend the network by changing and extending the input and output. For the input, we consider to use not only the board to play called *beforestate*[2] but also the board after the slides (and before appearing a new tile) called *afterstate*. For the output, we compute policy and/or value (of expected score until the game end).

Network Structure. Based on the network CNN12 in the first experiment, we design three additional networks as shown in Fig. 5. Note that the network Policy is the same as CNN12 in Fig. 2 (c).

The network Value is a straightforward conversion to a value network. It takes an afterstate of the intended move and computes a single value as the output. The network Dual is a combination of Policy and Value, where the parameters are mostly shared in the network. We expected that by this design with shared variables, we could capture features in the boards more effectively. The last network Policy AS is also a policy network but it takes not only the beforestate but also four afterstates corresponding to all the moves. Parameters are shared among five subnetworks. It is worth noting that in Game 2048 an afterstate

---

[2] The terms beforestate and afterstate are from Szubert and Jaśkowski [12].

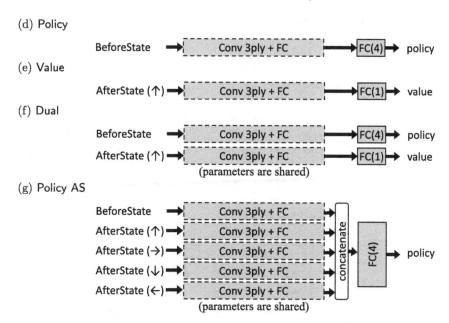

Fig. 5. Structure of networks used in the second experiment. The dashed parts (Conv 3ply + FC) correspond to one in Fig. 2 (c).

Table 6. Size and cost of neural networks used in the second experiment.

| Name | # parameters | Learning | | Playing | |
		Time	GPU memory	Time	GPU memory
Policy	3 681 284	117 min	3 181 MiB	1.95 ms	765 MiB
Value	3 680 513	136 min	3 581 MiB	3.33 ms	765 MiB
Dual (policy)	3 681 541	248 min	3 581 MiB	3.70 ms	761 MiB
(value)				5.45 ms	765 MiB
Policy AS	3 685 380	370 min	4 733 MiB	6.89 ms	765 MiB

could completely differ from the beforestate. We expect that the performance of the player would be improved by extending the input with afterstates.

Supervised Learning. From the results of the first experiment, we use the dataset acg17 only in the second experiment.

The loss function for the Policy and Policy AS networks is cross entropy (not changed). The loss function for the Value network is mean square error against the score obtained until the game end. In the Dual network, we have two loss values: the cross entropy of the policy part (*policy loss*) and the mean square error of the value part (*value loss*). Since the scale of the two loss values is very different, we scale the value loss down to the policy loss based on the ratio in the previous batch, before summing up the two loss values.

Table 7. Evaluation results of the second experiment.

Name		Loss	Agreement [%]	Score
Policy	1×10^8 actions	0.535 ± 0.018	75.63 ± 0.63	$77\,074 \pm\ 6\,351$
	2×10^8 actions	0.513 ± 0.027	75.92 ± 0.63	$91\,728 \pm\ 5\,609$
Value	1×10^8 actions	$2.27 \times 10^{10} \pm 3.9 \times 10^8$	32.77 ± 0.34	$1\,091 \pm\quad 190$
	2×10^8 actions	$2.26 \times 10^{10} \pm 3.9 \times 10^8$	29.82 ± 0.12	$1\,121 \pm\quad 199$
Dual (policy)	1×10^8 actions	0.127 ± 0.023	23.74 ± 0.66	$1\,431 \pm\quad 162$
	2×10^8 actions	0.120 ± 0.017	23.86 ± 0.33	$1\,209 \pm\quad 281$
(value)	1×10^8 actions	$2.28 \times 10^{10} \pm 2.4 \times 10^8$	32.57 ± 0.37	$3\,000 \pm\quad 210$
	2×10^8 actions	$2.27 \times 10^{10} \pm 3.8 \times 10^8$	32.90 ± 0.10	$3\,021 \pm\quad 451$
Policy AS	1×10^8 actions	0.514 ± 0.021	77.01 ± 0.33	$174\,486 \pm\ 7\,407$
	2×10^8 actions	0.481 ± 0.025	78.01 ± 0.08	$215\,803 \pm 12\,060$

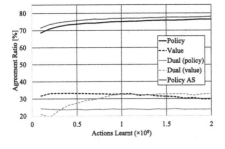

Fig. 6. Agreement ratio of the second experiment.

Fig. 7. Average score of the second experiment.

The other settings (number of actions learnt, methods of evaluation, playing method by majority vote, etc.) are the same as the first experiment.

Table 6 summarizes the number of parameters, time and GPU memory for the learning and playing phases. By the sharing of parameters, the numbers of parameters of the four networks are almost the same. For the GPU memory for the training phase, Policy AS consumed a bit more than the others. Whereas, the training time and the playing time differ considerably. Playing with value networks (Value and Dual) took longer time than policy networks, because we need to compute values for four afterstates. The network Policy AS took the longest time[3] both for training and playing because of the five paths for the beforestate and the afterstates.

Evaluation. Table 7 shows the experiment results after learning 1×10^8 and 2×10^8 actions. Figures 6 and 7 plot the agreement ratio and the average score of test plays, respectively. Dual (policy) and Dual (value) means that we used the policy part and the value part for the evaluation, respectively.

---

[3] But the factor was less than 3.6.

First of all, the obtained networks for Value and Dual completely failed to play, even though the loss values decreased over the training. Due to the randomness in Game 2048, scores after a certain board could change by more than 10^6. Considering this fact, the loss values about 2.26×10^{10} were sufficiently small. However, the scores of test plays with these two value networks were at the level of a random player. This bad result coincides with our previous trials of supervised learning for value networks [3]. We guess that the large variance of scores spoiled the training of value networks, and detailed analysis remains as our future work.

For the network Dual, the loss value of the policy part decreased to 0.12, which was much smaller than that of the network Policy. On the contrary, the agreement ratio and the score of test plays were quite bad. Since the loss values and the results of evaluations were inconsistent, we checked the output of the policy part and found that the outputs were either zero or one for most of the cases. Therefore, we guess that it started overfitting at an early stage of the training.

We achieved a success only with the network Policy AS. Though the improvement in terms of the loss value and the agreement ratio was rather small, we obtained a large improvement in terms of score of test plays. By attaching afterstates to the input, the network could look-ahead by a move and thus avoided to select some dangerous states. This reason was supported by the clear rate (achieving a 2048-tile) in the test plays: the clear ratio of the Policy network was 87.2% and that of the Policy AS network increased up to 98.8%.

Note that the average score of the acg17 players with the greedy (1-ply search) plays was 211 017. Our Policy AS player is the first neural-network player that achieved almost the same performance as top-level N-tuple-network players by a greedy (1-ply search) play if we ignore the playing time[4].

6   Conclusion

In this paper, we further investigated the NN-based players in two aspects. Firstly, we focused on the component of the networks. We compared NN-based players with full-connect layers, convolution layers with 2×2 filters, and convolution layers with 1×2 filters. We also compared the performance of these players by using two sets of playlogs from two different players. Secondly, we explored the input and/or output of networks. We designed four networks by changing the input and output of the networks.

The following are the important findings from experiment results. (1) Using full-connect layers only (multi-layer perceptrons) was not as good as convolutional networks with small filters. (2) Convolution layers with small 1×2 filters was the best for Game 2048. (3) We could obtain very different players from playlogs with similar performance. (4) Training of value networks failed for Game 2048. (5) Performance of policy networks was improved largely by extending the

---

[4] The playing time of player Policy AS was about 5 000 times as long as that of N-tuple-network players.

input with afterstates. (6) The best average score achieved was 215 803 without search techniques.

One important topic for future work is to identify the reason of the failure for the value networks. Another topic is to apply reinforcement learning methods to the networks to obtain better results.

Acknowledgment. The training data `acg17` and `nneo` used in this study were generated with the support of the IACP cluster in Kochi University of Technology.

References

1. Cirulli, G.: 2048 (2014). http://gabrielecirulli.github.io/2048/
2. David, O.E., Netanyahu, N.S., Wolf, L.: DeepChess: end-to-end deep neural network for automatic learning in chess. In: Villa, A.E.P., Masulli, P., Pons Rivero, A.J. (eds.) ICANN 2016, Part II. LNCS, vol. 9887, pp. 88–96. Springer, Cham (2016). https://doi.org/10.1007/978-3-319-44781-0_11
3. Fujita, R., Matsuzaki, K.: Improving 2048 player with supervised learning. In: Proceedings of 6th International Symposium on Frontier Technology, pp. 353–357 (2017)
4. Guei, H., Wei, T., Huang, J.B., Wu, I.C.: An early attempt at applying deep reinforcement learning to the game 2048. In: Workshop on Neural Networks in Games (2016)
5. Jaśkowski, W.: Mastering 2048 with delayed temporal coherence learning, multi-stage weight promotion, redundant encoding and carousel shaping. IEEE Trans. Comput. Intell. AI Games **10**(1), 3–14 (2018)
6. Kondo, N., Matsuzaki, K.: Playing game 2048 with deep convolutional neural networks trained by supervised learning. J. Inf. Process. **27**, 340–347 (2019)
7. Lai, M.: Giraffe: Using Deep Reinforcement Learning to Play Chess. Master's thesis, Imperial College London (2015). arXiv:1509.01549v1
8. Matsuzaki, K.: Developing a 2048 player with backward temporal coherence learning and restart. In: Winands, M.H.M., van den Herik, H.J., Kosters, W.A. (eds.) ACG 2017. LNCS, vol. 10664, pp. 176–187. Springer, Cham (2017). https://doi.org/10.1007/978-3-319-71649-7_15
9. Samir, M.: An attempt at applying Deep RL on the board game 2048 (2017). https://github.com/Mostafa-Samir/2048-RL-DRQN
10. Silver, D., et al.: Mastering Chess and Shogi by self-play with a general reinforcement learning algorithm. arXiv 1712.01815 (2017)
11. Silver, D.: Mastering the game of go without human knowledge. Nature **550**, 354–359 (2017)
12. Szubert, M., Jaśkowski, W.: Temporal difference learning of N-tuple networks for the game 2048. In: 2014 IEEE Conference on Computational Intelligence and Games, pp. 1–8 (2014)
13. tjwei: A deep learning AI for 2048 (2016). https://github.com/tjwei/2048-NN
14. Virdee, N.: Trained a convolutional neural network to play 2048 using deep-reinforcement learning (2018). https://github.com/navjindervirdee/2048-deep-reinforcement-learning
15. Yeh, K.-H., Wu, I.-C., Hsueh, C.-H., Chang, C.-C., Liang, C.-C., Chiang, H.: Multi-stage temporal difference learning for 2048-like games. IEEE Trans. Comput. Intell. AI Games **9**(4), 369–380 (2016)

16. Wiese, G.: 2048 reinforcement learning (2018). https://github.com/georgwiese/2048-rl
17. Xiao, R.: nneonneo/2048-ai (2015). https://github.com/nneonneo/2048-ai

A Novel Application for Game Tree Search - Exploiting Pruning Mechanisms for Quantified Integer Programs

Michael Hartisch$^{(\boxtimes)}$ and Ulf Lorenz

Chair of Technology Management, University of Siegen, Siegen, Germany
{michael.hartisch,ulf.lorenz}@uni-siegen.de

Abstract. We investigate pruning in search trees of so-called quantified integer (linear) programs (QIPs). QIPs consist of a set of linear inequalities and a minimax objective function, where some variables are existentially and others are universally quantified. A good way to solve a QIP is to apply game tree search, enhanced with non-chronological backjumping. We develop and theoretically substantiate tree pruning techniques based upon algebraic properties. The presented Strategic Copy-Pruning mechanism allows to *implicitly* deduce the existence of a strategy in linear time (by static examination of the QIP-matrix) without explicitly traversing the strategy itself. We show that the implementation of our findings can massively speed up the search process.

1 Introduction

Prominent solution paradigms for optimization under uncertainty are Stochastic Programming [5], Robust Optimization [3], Dynamic Programming [2], Sampling [11] and POMDP [21]. Relatively unexplored are the abilities of linear programming extensions for PSPACE-complete problems. In the early 2000s the idea of universally quantified variables, as they are used in quantified constraint satisfaction problems [10], was picked up again [27], coining the term quantified integer program (QIP). Quantified integer programming is a direct, very formal extension of integer linear programming (IP), making QIPs applicable in a very natural way. They allow robust multistage optimization extending the two-stage approach of Robust Optimization [3]. Multistage models - in contrast to two-stage models - allow more precise planning strategies as uncertain events typically do not occur all at the same time (delay in timetables, changed cost estimate for edges in a graph, alternating moves in games).

Let us start with the following illustrative application. There are b runways at your airport and all arriving airplanes must be assigned to exactly one time slot for the landing (therefore a natural *worst-case* optimization problem). Further,

This research is partially supported by the German Research Foundation (DFG) project "Advanced algorithms and heuristics for solving quantified mixed - integer linear programs".

T. Cazenave et al. (Eds.): ACG 2019, LNCS 12516, pp. 66–78, 2020.
https://doi.org/10.1007/978-3-030-65883-0_6

the airplanes are expected to arrive within some time window and hence the assigned time slot must adhere to those time windows. Finding an initial matching, even an optimal one considering some objective function, can be modeled and solved using mixed integer programming techniques [14]. However, the time windows are uncertain due to adjusted airspeed (due to weather) or operational problems and an initial schedule might become invalid (see for example Fig. 1). Thus, one is interested in a robust initial plan that can be adapted cheaply,

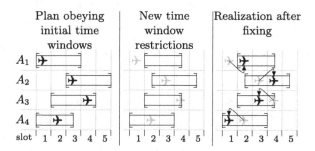

Fig. 1. Process of runway scheduling: A schedule for the initial time windows is made (left). If the predicted time windows differ from the actually occurring time windows (middle), the initial plan becomes invalid and a new scheduling must be found (right).

e.g. the initial and adapted time slot of each airplane should not be too far apart from each other [13]. These uncertain events, however, do not uncover all at the same time: final time slots must be assigned to some airplanes while for other airplanes the actual time window is still unknown. This problem is literally crying out to be modeled as a QIP.

A solution of a QIP is a strategy – in the game tree search sense [22] – for assigning existentially quantified variables such that some linear constraint system is fulfilled. By adding a minimax objective function the aim is to find the best strategy [19]. As not unusual in the context of optimization under uncertainty [3,4] a polyhedral uncertainty set can be used [12]. There are two different ways known how to tackle a QIP: On the one hand the so-called deterministic equivalent program can be built, similar to the ones known from stochastic programming [5], and solved using standard integer programming solvers. On the other hand the more natural approach is to conduct a game tree search [18,26]. We are interested in utilizing game solving techniques [25,28] in combination with linear programming techniques as well as pruning and backjumping techniques from QBF [6]. Recently our solver for quantified mixed integer programs was made available as open source. This solver combines techniques known from game tree search, linear programming and QBF [9].

An optimization task is often split up into two parts: finding the optimal solution itself and proving that no better solution can exist. For the latter, it turned out that applying backjumping techniques as utilized by QBF-solvers [29] and cutting planes as commonly used in integer programming [20] are also highly beneficial for QIPs in order to assess that no (better) strategy can exist in certain subtrees. For the first task, however, it seems that the exponential number of leaves belonging to a strategy must be traversed explicitly. This is certainly true in the worst-case. However, typically there are "difficult" parts of a game tree where a very deliberated substrategy must be found but also other parts where a less sophisticated substrategy suffices. In this paper we present a procedure, called strategic copy-pruning (SCP), that is capable of recognizing such subtrees making it possible to *implicitly* deduce the existence of a winning strategy therein. In contrast to similar ideas in QBF, as e.g. counterexample guided abstraction refinement [16], an optimization process over a minimax objective must be considered. Further, our SCP draws its power not from memory-intensive learning, but from deep findings in the search tree. This perspective has led to remarkable achievements in the past [7,15]. For game tree search there are already several algorithms trying to rapidly show the existence of winning strategies such as Kawano's simulation [17], MTD(f) [23] and (nega)scout [24]. They, however, always have to traverse an exponential number of leafs. In our experiments, SCP often allows to conclude the existence of an winning strategy with a linear number of algebraic operations and in particular, in those cases it is not necessary to examine an exponential number of leaves resulting in a significant performance improvement both in time (about a factor 4) and number of solved instances. The effect of SCP is reinforced if the sequence of variable assignments predicted as optimal by minimax for both sides, called the principal variation [8], is traversed in an early stage of the tree search. Detecting and verifying this particular variable assignment is essential in order to obtain the objective value. Thus having reasonable knowledge of which universal variable assignments are particularly vicious can massively boost the search process. Several heuristics exist to analyze and find such promising moves in a game tree search environment [1,23,25]. The paper is organized as follows: First basic definitions and notations regarding QIPs are presented. Then two pruning techniques for the QIP game tree search are introduced: First, the well known monotonicity [6] of variables is recaptured. Second, as our main result, we derive from already found strategies the existence of winning strategies in other branches. This happens in a way such that these branches do not need to be investigated explicitly. Finally the conducted experiments are presented.

2   Preliminaries: Basics of Quantified Integer Programming

Let $n \in \mathbb{N}$ be the number of variables and $x = (x_1, \ldots, x_n)^\top \in \mathbb{Z}^n$ a vector of variables.[1] For each variable x_j its domain $\mathcal{L}_j$ with $l_j, u_j \in \mathbb{Z}$, $l_j \leq u_j$, $1 \leq j \leq n$, is given by $\mathcal{L}_j = \{y \in \mathbb{Z} \mid l_j \leq y \leq u_j\}$. The domain of the entire variable vector is described by $\mathcal{L} = \{y \in \mathbb{Z}^n \mid \forall j \in \{1, \ldots, n\} : y_j \in \mathcal{L}_j\}$,

---

[1] $\mathbb{Z}$, $\mathbb{N}$ and $\mathbb{Q}$ denote the set of integers, natural numbers, and rational numbers, respectively.

i.e. each variable must obey its domain. Let $Q \in \{\exists, \forall\}^n$ denote the vector of quantifiers. We call $\mathcal{E} = \{j \in \{1, \ldots, n\} \mid Q_j = \exists\}$ the set of existential variables and $\mathcal{A} = \{j \in \{1, \ldots, n\} \mid Q_j = \forall\}$ the set of universal variables. Further, each maximal consecutive subsequence in Q consisting of identical quantifiers is called *quantifier block* with $B_i \subseteq \{1, \ldots, n\}$ denoting the i-th block. Let $\beta \in \mathbb{N}$, $\beta \leq n$, denote the number of blocks and thus $\beta - 1$ is the number of quantifier changes. The variable vector of variable block B_i will be referred to as $x^{(i)}$.

Definition 1 (Quantified Integer Linear Program (QIP)). *Let $A \in \mathbb{Q}^{m \times n}$ and $b \in \mathbb{Q}^m$ for $m \in \mathbb{N}$ and let $\mathcal{L}$ and Q be given as described above. Let $c \in \mathbb{Q}^n$ be the vector of objective coefficients and let $c^{(i)}$ denote the vector of coefficients belonging to block B_i. Let the term $Q \circ x \in \mathcal{L}$ with the component wise binding operator $\circ$ denote the* quantification vector $(Q_1 x_1 \in \mathcal{L}_1, \ldots, Q_n x_n \in \mathcal{L}_n)$ *such that every quantifier Q_j binds the variables x_j to its domain $\mathcal{L}_j$. We call $(A, b, c, \mathcal{L}, Q)$ with*

$$z = \min_{B_1} \left(c^{(1)} x^{(1)} + \max_{B_2} \left(c^{(2)} x^{(2)} + \ldots \min_{B_\beta} c^{(\beta)} x^{(\beta)} \right) \right)$$

$$s.t. \ Q \circ x \in \mathcal{L} : \ Ax \leq b \tag{$\star$}$$

a QIP with objective function (for a minimizing existential player).

For simplicity's sake, and since it goes well with the example in Fig. 1, we will consider only binary QIPs, i.e. $l_j = 0$ and $u_j = 1$ for all $j \in \{1, \ldots, n\}$. However, note that our results and in particular Theorem 1 can easily be adapted to be valid for general integer variables.

A QIP instance can be interpreted as a two-person zero-sum game between an *existential player* setting the existentially quantified variables and a *universal player* setting the universally quantified variables with payoff z. The variables are set in consecutive order according to the variable sequence. Consequently, we say that a player makes the move $x_k = y$ if she fixes the variable x_k to $y \in \mathcal{L}_k$. At each such move, the corresponding player knows the settings of $x_1, \ldots, x_{k-1}$ before taking her decision x_k. If the completely assigned vector $x \in \mathcal{L}$ satisfies the linear constraint system $Ax \leq b$, the existential player pays $z = c^\top x$ to the universal player. If x does not satisfy $Ax \leq b$, we say *the existential player loses* and the payoff will be $+\infty$. This is a small deviation from conventional zero-sum games but using[2] $\infty + (-\infty) = 0$ also fits for zero-sum games. The chronological order of the variable blocks given by Q can be represented using a game tree $G = (V, E, c)$ with $V = V_\exists \cup V_\forall \cup V_L$ consisting of existential, universal and leaf nodes [9]. Thus, a path from the root to a leaf represents a play of the QIP and the sequence of edge labels encodes its moves, i.e. the corresponding variable assignments. Solutions of a QIP are strategies [9]. In the following, the word *strategy* will always refer to an *existential* strategy. A strategy is called a *winning strategy* if all paths from the root node to a leaf represent a vector x such that $Ax \leq b$. A QIP is called *feasible* if $(\star)$ is true (see Definition 1), i.e. if a

---

[2] This is only a matter of interpretation and consequences are not discussed further.

winning strategy exists. If there is more than one winning strategy, the objective function aims for a certain (the "best") one. The value of a strategy is given by its minimax value which is the maximum value at its leaves [22]. Note that a leaf not fulfilling $Ax \leq b$ can be represented by the value $+\infty$. The objective value of a feasible QIP is the minimax value at the root, i.e. the minimax value of the optimal winning strategy, defined by the *principal variation* (PV) [8]: the sequence of variable assignments being chosen during optimal play. For any $v \in V$ we call $f(v)$ the outcome of optimal play by both players starting at v.

Example. Let us consider a QIP with $n = 4$ binary variables:

$$
\begin{array}{lllll}
\min(\ 2x_1 & \max(\ -2x_2 & \min(\ -3x_3 & \max(\ -2x_4)))) \\
\text{s.t.} \ \exists x_1 \in \{0,1\} & \forall x_2 \in \{0,1\} & \exists x_3 \in \{0,1\} & \forall x_4 \in \{0,1\} & : \\
x_1 & & +x_3 & & \leq 2 \\
-x_1 & & +x_3 & -x_4 & \leq 0 \\
& -x_2 & +x_3 & -x_4 & \leq 0 \\
-x_1 & +x_2 & -x_3 & +x_4 & \leq 1
\end{array}
$$

The minimax value of the root node (for the minimizing starting player) of the game tree is 2 and the principal variation is given by $x_1 = 1$, $x_2 = 0$, $x_3 = 0$ and $x_4 = 0$. The inner node at level 1 resulting from setting $x_1 = 0$ has the minimax value $+\infty$, i.e. after setting $x_1 = 0$ there exists no winning strategy.

3   Pruning in QIP Search Trees

3.1   Theoretical Analysis

In a natural way, a quantified integer program can be solved via game tree search. During such a tree search we are interested in quickly evaluating or estimating the minimax value of nodes, i.e. we want to examine the optimal (existential) strategy of the corresponding subtree. In order to speed up the search process, limiting the number of subtrees that need to be explored is extremely beneficial. Such pruning operations are applied in many search based algorithms, e.g. the alpha-beta algorithm [18], branch-and-bound [20] and DPLL [29]. In the following, we will present two approaches that allow pruning in a QIP game tree search, and thus in a strategic optimization task.

In case of QIPs certain variable assignments never need to be checked as they are worse than their counterparts. The concept of monotone variables is already well known for quantified boolean formulas [6] and integer programming [20].

Definition 2 (Monotone Variable)
A variable x_k of a QIP is called monotone if it occurs with only positive or only negative sign in the matrix and objective, i.e. if the entries of A and c belonging to x_k are either all non-negative or all non-positive.

Using this easily verifiable monotonicity allows us to omit certain subtrees a priori since solving the subtree of its sibling is guaranteed to yield the desired minimax value.

In contrast to this usage of prior knowledge we also want to gather *deep knowledge* during the search process: found strategies in certain subtrees can be useful in order to assess the minimax value of related subtrees rapidly. The idea is based upon the observation that typically in only a rather small part of the game tree a distinct and crafty strategy is required in order to ensure the fulfillment of the constraint system: in the right-hand side subtree of Fig. 2 it suffices to find a fulfilling existential variable assignment for only one scenario (universal variable assignment) and reuse it in the other branches.

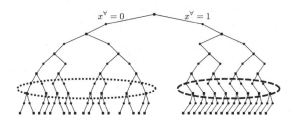

Fig. 2. Illustrative strategy for which the universal assignment $x^\forall = 1$ entails a simple winning strategy: Regardless of future universal decisions existential variables can be set in a certain simple way, e.g. the existential decisions in the dashed ellipse are all the same. $x^\forall = 0$ on the other hand compels a more clever strategy, e.g. the existential decisions in the dotted ellipse differ depending on previous universal decisions.

Theorem 1. *[Strategic Copy-Pruning (SCP)]*
Let $k \in \mathcal{A}$ and let $(\tilde{x}_1, \ldots, \tilde{x}_{k-1}) \in \{0,1\}^{k-1}$ be a fixed variable assignment of the variables $x_1, \ldots, x_{k-1}$. Let $v \in V_\forall$ be the corresponding universal node in the game tree. Let $\tilde{w} \in V$ and $\hat{w} \in V$ be the two children of v corresponding to the variable assignment $\tilde{x}_k$ and $\hat{x}_k = 1 - \tilde{x}_k$ of the universal variable x_k, respectively. Let there be an optimal winning strategy for the subtree below $\tilde{w}$ with minimax value $f(\tilde{w}) = \tilde{z}$ defined by the variable assignment $\tilde{x} = (\tilde{x}_1, \ldots, \tilde{x}_n) \in \{0,1\}^n$, i.e. $\tilde{z} = c^\top \tilde{x}$. If the minimax value of the copied strategy for the subtree below $\hat{w}$ - obtained by adoption of future[3] existential variable assignments as in $\tilde{x}$ - is not larger than $\tilde{z}$ and if this copied strategy constitutes a winning strategy then $f(v) = \tilde{z}$. Formally: If both

$$c_k(\hat{x}_k - \tilde{x}_k) + \sum_{\substack{j \in \mathcal{A}, \ j > k \\ \text{and } c_j \geq 0}} c_j(1 - \tilde{x}_j) - \sum_{\substack{j \in \mathcal{A}, \ j > k \\ \text{and } c_j < 0}} c_j \tilde{x}_j \leq 0 \qquad (1)$$

and

$$\sum_{\substack{j \in \mathcal{E} \\ \text{or } j < k}} A_{i,j} \tilde{x}_j + A_{i,k} \hat{x}_k + \sum_{\substack{j \in \mathcal{A}, \ j > k \\ \text{and } A_{i,j} > 0}} A_{i,j} \leq b_i \qquad (2)$$

---

[3] *Future* means variable blocks with index $\geq k$.

for all constraints $i \in \{1, \ldots, m\}$ then $f(v) = \tilde{z}$.

For clarification note that Condition (1) ensures that the change in the minimax value of the copied strategy, resulting from flipping x_k and using the worst case assignment of the remaining future universal variables, is not positive, i.e. that its minimax value is still smaller than or equal to $\tilde{z}$. Condition (2) verifies that every constraint is satisfied in each leaf of the copied strategy by ensuring the fulfillment of each constraint in its specific worst case scenario.

Proof. If (2) is satisfied there automatically exists a winning strategy for the subtree of v corresponding to $x_k = \hat{x}_k$ with root node $\hat{w}$, since for any future universal variable assignment the assignment of upcoming existential variables as in $\tilde{x}$ fulfills the constraint system. Further, the minimax value $\hat{z}$ of this strategy is smaller than or equal to $\tilde{z}$ due to Condition (1):

$$\hat{z} = \sum_{\substack{j \in \mathcal{E} \\ \text{or } j < k}} c_j \tilde{x}_j + c_k \hat{x}_k + \sum_{\substack{j \in \mathcal{A},\, j > k \\ \text{and } c_j \geq 0}} c_j$$

$$\overset{(1)}{\leq} \sum_{\substack{j \in \mathcal{E} \\ \text{or } j < k}} c_j \tilde{x}_j + c_k \hat{x}_k + \sum_{\substack{j \in \mathcal{A},\, j > k}} c_j \tilde{x}_j \quad = \tilde{z}$$

Hence, the (still unknown) optimal strategy for the subtree below $\hat{w}$ has a minimax value smaller than or equal to $\tilde{z}$, i.e. $f(\hat{w}) \leq \hat{z} \leq \tilde{z} = f(\tilde{w})$. Therefore, $f(v) = f(\tilde{w}) = \tilde{z}$.

Note that, since $A\tilde{x} \leq b$, Condition (2) is trivially fulfilled for any constraint $i \in \{1, \ldots, m\}$ with $A_{i,j} = 0$ for all $j \in \mathcal{A}, j \geq k$, i.e. constraints that are not influenced by future universal variables do not need to be examined. Hence, only a limited number of constraints need to be checked in case of a sparse matrix. Further, note that (1) is fulfilled if $c_j = 0$ for all $j \in \mathcal{A}, j \geq k$, i.e. if the future universal variables have no direct effect on the objective value. In particular, if $c = 0$, i.e. it is a satisfiabilty problem rather than an optimization problem, Condition (1) can be neglected as it is always fulfilled.

3.2   SCP Implementation Details

As soon as a leaf v is found during the tree search with the corresponding x_v being a potentially new PV for this subtree the following mechanism is invoked: the two Conditions (1) and (2) of Theorem 1 are checked at each universal node starting from this leaf towards the root (Line 5). While both conditions are fulfilled the corresponding universal nodes are marked as potentially finished. If one of the conditions is not satisfied the remaining universal nodes above are marked as unfinished. If a level is closed during the tree search and the above universal node is marked as potentially finished this level also can be closed immediately as a strategy is guaranteed in the other branch with worse objective value (from the universal player's point of view). The unmarking of universal nodes (Line 8) is necessary since Theorem 1 demands x_v to be the actual PV of this subtree and hence previous markings where made based on a false assumption.

Algorithm 1. Marking of potentially finished universal nodes

Input: leaf node v

  1: useSCP=true;
  2: **repeat**
  3:    v=parent(v);
  4:    **if** $v \in V_\forall$ **then**
  5:        **if** useSCP **and** v fulfills Conditions (1) and (2) **then**
  6:            mark v as potentially finished;
  7:        **else**
  8:            useSCP=false; mark v as unfinished;
  9:        **end if**
10:    **end if**
11: **until** v is root node

3.3 Example

Consider the following binary QIP (The min/max alternation in the objective and the binary variable domains are omitted):

$$\min 2x_1 + 3x_2 - 2x_3 - 2x_4 + x_5$$
$$\text{s.t. } \exists x_1 \quad \forall x_2 \quad \exists x_3 \quad \forall x_4 \quad \exists x_5 \quad :$$
$$x_1 - x_2 + x_3 + 3x_4 - x_5 \leq 2$$
$$3x_1 + 2x_2 + 3x_3 + x_4 - 2x_5 \leq 1$$

Starting at the root node of the corresponding game tree we can immediately omit the subtree corresponding to $x_1 = 1$ due to the monotonicity of x_1. Keep in mind that the result of Theorem 1 is particularly beneficial if the search process of a QIP solver first examines the principal variation, i.e. the variable assignment defining the actual minimax value. Assume the search process follows the path drawn thick in Fig. 3 to node v_8, i.e. the path corresponding to the variable assignment $x_1 = 0$, $x_2 = 1$, $x_3 = 0$ and $x_4 = 0$. Setting $x_5 = 1$ is optimal in this case, as $x_5 = 0$ would violate the second constraint. Hence, the minimax value of v_8 is 4. On the way up in the search tree we then want to determine $f(v_5)$. As (1) and (2) are fulfilled for $k = 4$, $\tilde{z} = 4$ and $\tilde{x} = (0, 1, 0, 0, 1)$ we know that $f(v_5) = 4$. That means we have (easily) verified a winning strategy starting from v_9 with minimax value smaller than or equal to 4. In node v_3 setting $x_3 = 1$ is obviously to the detriment of the existential player, as the second constraint would become unfulfillable. Hence, $f(v_3) = f(v_5) = 4$. In node v_1 we once again try to apply Theorem 1 by copying the existential decisions of x_3 and x_5 in the thick path to the not yet investigated subtree associated with $x_2 = 0$. As (1) and (2) are fulfilled for $k = 2$, $\tilde{z} = 4$ and $\tilde{x} = (0, 1, 0, 0, 1)$ this attempt is successful and $f(v_1) = 4$. Note that by applying Theorem 1 the minimax value of the subtrees below v_2 and v_9 are not known exactly: in particular we only obtain $f(v_2) \leq \hat{z} = 1$, whereas a better strategy exists resulting in $f(v_2) = 0$ (Setting $x_5 = 0$ in node v_6).

Hence, by finding the principal variation first (thick path), exploiting monotonicity of x_1 at node v_0, Theorem 1 at node v_1 and v_5 and some further reasoning

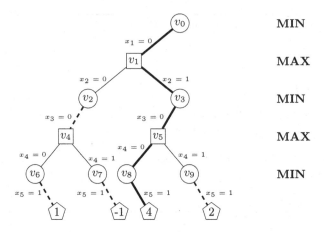

Fig. 3. Optimal winning strategy for the QIP. Circular nodes are existential decision nodes, rectangular nodes are universal decision nodes and pentagonal nodes are leaves. The values given in the leaves constitute the objective value corresponding to the variable assignment along the path from the root to this leaf. The dashed lines indicate that those existential decisions where simply copied from the path drawn thicker.

from linear programming at node v_3 and v_8 the minimax value at the root node v_0 was found to be 4 with optimal first stage solution $x_1 = 0$.

Theorem 1 can particularly come into effect if the branching decisions at universal nodes result in rather vicious scenarios, i.e. in variable assignments restricting the constraint system and maximizing the objective value. Hence, the applicability of the presented results largely depends on the implemented diving and sorting heuristic.

4   Solver, Experiments and Results

We use our open source[4] solver Yasol [9] to analyze the theoretical findings. The solver basically performs an enhanced alpha-beta-search and proceeds in two phases: a *feasibility phase*, where it is checked whether the instance has any solution at all, and an *optimization phase* for finding the provable optimal solution. We enhanced this solver in two different ways:

1. The detection and exploitation of monotone variables.
2. The adoption of existing winning strategies from one branch of a universal node to another (SCP).

The SCP-enhancement can be switched on and off in both phases separately.

The instances used to study the effect of the presented results are runway scheduling problems under uncertainty as motivated in the introduction.

---

[4] Sources are available at http://www.q-mip.org.

They were created following the ideas presented in [13]. The task is to find a b-matching: all airplanes must be assigned to exactly one time slot, while one time slot can take in at most b airplanes. Furthermore, the airplanes must land within an uncertain time window. Hence, we are interested in an initial matching plan that can be fixed cheaply if the mandatory time windows for some airplanes do not contain the initially scheduled time slot. The testset contains 29 instances[5], varying in the number of airplanes, the number of time slots, the type of allowed disturbances, the number of universal blocks and the cost function. In terms of the sizes of the (solved feasible) instances this results in between 100–300 existential variables, 10–30 universal variables and 50–100 constraints.

Table 1. Number of solved instances dependend on the solver setting: exploitation of monotone variables and SCP in different phases.

monotonicity	SCP	solved instances
off	both phases	24
	only optimization phase	21
	only feasibility phase	16
	off	14
on	both phases	25
	only optimization phase	25
	only feasibility phase	24
	off	23

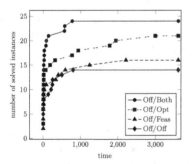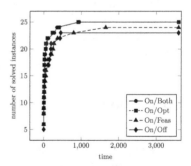

Fig. 4. Comparison of the effect of SCP in different phases of the solver without exploiting monotonicity (left) and when also exploiting monotonicity (right).

In Table 1 and Fig. 4 the number of solved instances and the cumulative solution diagram is displayed for different settings. For each instance a maximum

---

[5] The studied benchmark instances and a brief explanation can be found at http://www.q-mip.org/index.php?id=41.

Table 2. Average time needed for the 23 instances solved in all four settings while exploiting monotonicity.

SCP setting	off	only feas	only opt	both
average runtime	84s	102s	25s	32s

of one hour solution time was provided. All experiments were executed on a PC with an Intel i7-4790 (3.6 GHz) processor and 32 GB RAM. If neither of the presented procedures is used 14 out of 29 instances are solved. Without taking advantage of the monotonicity SCP can be beneficial in either solution phase regarding the number of solved instances. If applied in both phases the number of solved instances is increased up to 24. When also exploiting the monotonicity the number of solved instances increases to 25. However, SCP turns out to be somewhat disadvantageous in the feasibility phase. Even though an additional instance is solved (24) compared to the setting with SCP turned off (23) the average solution time increases: in Table 2 the average time needed for the 23 instances solved by all versions with turned on monotonicity is displayed. Additionally using SCP in the feasibility phase slightly increases the average solution time. Our conjecture is that this is due to biasing effects. Four instances with more than 100 universal variables and 10000 existential variables were not solved at all. However, there also are infeasible instances of the same magnitude that are solved within seconds. In order to assess the performance results, we also built the deterministic equivalent program of each instance and tried to solve the resulting integer program using CPLEX 12.6.1.0, a standard MIP solver. Only six of the 29 instances where solved this way, given the same amount of time, while for 14 instances not even the construction of the corresponding DEP could be finished, some of them because of the limited memory of 32 GB RAM. Experiments conducted on a QBF test collection of 797 instances, taken from www.qbflib.org, also show positive effects for the SCP version. When only exploiting monotonicity 644 instances are solved. If additionally SCP is turned on 674 instances can be solved. Further, the solution time decreases by 15%.

5   Conclusion

We introduced the concept of strategic copy-pruning (SCP) during tree search for quantified integer programs. SCP makes it possible to omit certain subtrees during the search process by implicitly verifying the existence of a strategy in linear time: finding a single leaf and applying SCP can be sufficient to guarantee an optimal strategy in an entire subtree. This is standing in contrast to existing algorithms in which the existence of a strategy is proven by traversing it explicitly. In addition, we presented how those findings can be applied in a search environment. Experiments showed that utilizing our approach resulted in a massive boost in both the number of solved instances and the solution time (about 4 times faster) on a particular testset. The achievement opens the door to solving larger and more complex real-world problems.

References

1. Akl, S., Newborn, M.: The principal continuation and the killer heuristic. In: ACM 1977, pp. 466–473 (1977)
2. Bellman, R.: Dynamic Programming. Dover Publications Incorporated, Mineola (2003)
3. Ben-Tal, A., Ghaoui, L.E., Nemirovski, A.: Robust Optimization. Princeton University Press, Princeton (2009)
4. Bertsimas, D., Brown, D., Caramanis, C.: Theory and applications of robust optimization. SIAM Rev. **53**(3), 464–501 (2011)
5. Birge, J.R., Louveaux, F.: Introduction to Stochastic Programming. Springer, New York (2011). https://doi.org/10.1007/978-1-4614-0237-4
6. Cadoli, M., Schaerf, M., Giovanardi, A., Giovanardi, M.: An algorithm to evaluate quantified Boolean formulae and its experimental evaluation. J. Autom. Reasoning **28**(2), 101–142 (2002)
7. Campbell, M., Hoane, A., Hsu, F.H.: Search control methods in deep blue. In: AAAI Spring Symposium on Search Techniques for Problem Solving under Uncertainty and Incomplete Information, pp. 19–23 (1999)
8. Campbell, M., Marsland, T.: A comparison of minimax tree search algorithms. Artif. Intell. **20**(4), 347–367 (1983)
9. Ederer, T., Hartisch, M., Lorenz, U., Opfer, T., Wolf, J.: Yasol: an open source solver for quantified mixed integer programs. In: Winands, M.H.M., van den Herik, H.J., Kosters, W.A. (eds.) ACG 2017. LNCS, vol. 10664, pp. 224–233. Springer, Cham (2017). https://doi.org/10.1007/978-3-319-71649-7_19
10. Gerber, R., Pugh, W., Saksena, M.: Parametric dispatching of hard real-time tasks. IEEE Trans. Comput. **44**(3), 471–479 (1995)
11. Gupta, A., Pál, M., Ravi, R., Sinha, A.: Boosted sampling: approximation algorithms for stochastic optimization. In: ACM 2004, pp. 417–426. ACM (2004)
12. Hartisch, M., Ederer, T., Lorenz, U., Wolf, J.: Quantified integer programs with polyhedral uncertainty set. In: Plaat, A., Kosters, W., van den Herik, J. (eds.) CG 2016. LNCS, vol. 10068, pp. 156–166. Springer, Cham (2016). https://doi.org/10.1007/978-3-319-50935-8_15
13. Heidt, A., Helmke, H., Kapolke, M., Liers, F., Martin, A.: Robust runway scheduling under uncertain conditions. JATM **56**, 28–37 (2016)
14. Helmke, H.: Scheduling algorithms for ATM applications–tools and toys. In: 2011 IEEE/AIAA 30th Digital Avionics Systems Conference, p. 3C2-1. IEEE (2011)
15. van den Herik, H., Nunn, J., Levy, D.: Adams outclassed by hydra. ICGA J. **28**(2), 107–110 (2005)
16. Janota, M., Klieber, W., Marques-Silva, J., Clarke, E.: Solving QBF with counterexample guided refinement. Artif. Intell. **234**, 1–25 (2016)
17. Kawano, Y.: Using similar positions to search game trees. Games No Chance **29**, 193–202 (1996)
18. Knuth, D., Moore, R.: An analysis of alpha-beta pruning. Artif. Intell. **6**(4), 293–326 (1975)
19. Lorenz, U., Wolf, J.: Solving multistage quantified linear optimization problems with the alpha-beta nested benders decomposition. EURO J. Comput. Optim. **3**(4), 349–370 (2015)
20. Nemhauser, G., Wolsey, L.: Integer and Combinatorial Optimization. Wiley-Interscience, New York (1988)

21. Nguyen, D., Kumar, A., Lau, H.: Collective multiagent sequential decision making under uncertainty. In: AAAI 2017. AAAI Press (2017)
22. Pijls, W., de Bruin, A.: Game tree algorithms and solution trees. Theoret. Comput. Sci. **252**(1), 197–215 (2001)
23. Plaat, A., Schaeffer, J., Pijls, W., de Bruin, A.: Best-first fixed-depth minimax algorithms. Artif. Intell. **87**(1–2), 255–293 (1996)
24. Reinefeld, A.: An improvement to the scout tree search algorithm. ICGA J. **6**(4), 4–14 (1983)
25. Schaeffer, J.: The history heuristic and alpha-beta search enhancements in practice. IEEE Trans. Pattern Anal. Mach. Intell. **11**(11), 1203–1212 (1989)
26. Silver, D., et al.: Mastering the game of go with deep neural networks and tree search. Nature **529**, 484–503 (2016)
27. Subramani, K.: Analyzing selected quantified integer programs. In: Basin, D., Rusinowitch, M. (eds.) IJCAR 2004. LNCS (LNAI), vol. 3097, pp. 342–356. Springer, Heidelberg (2004). https://doi.org/10.1007/978-3-540-25984-8_26
28. Winands, M., van den Herik, H., Uiterwijk, J., van der Werf, E.: Enhanced forward pruning. Inf. Sci. **175**(4), 315–329 (2005)
29. Zhang, L.: Searching for truth: techniques for satisfiability of Boolean formulas. Ph.D. thesis, Princeton, USA (2003)

New Hex Patterns for Fill and Prune

Nicolas Fabiano[1] and Ryan Hayward[2(✉)]

[1] Dèpartement d'Informatique, ENS Ulm, Paris, France
[2] Department of Computing Science, University of Alberta, Edmonton, Canada

Abstract. For a position in the game of Hex, a fill pattern is a sub-position with one or more empty cells that can be filled without changing the position's minimax value. A cell is prunable if it can be ignored when searching for a winning move. We introduce two new kinds of Hex fill – mutual and near-dead – and some resulting fill patterns; we show four new permanently-inferior fill patterns; and we present three new prune results, based on strong-reversing, reversing, and game-history respectively. Experiments show these results slightly reducing solving time on 8×8 openings.

1 Introduction

Black and White alternate turns; on each turn, a player places one stone of their color on any unoccupied cell. The winner is the player who forms a chain of their stones connecting their two opposing board sides. See Fig. 1.

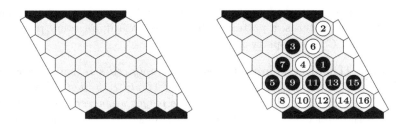

Fig. 1. An empty 5 × 5 Hex board and a game won by White.

Hex never ends in a draw [3] and it is never disadvantageous to move [14], so by strategy-stealing the first player has a winning strategy on an empty board [2]. But, for general sizes, no explicit first-player-wins strategy is known, nor is the location of a winning opening move. For boards up to 9×9 (resp. the 10×10 board) all (4) winning opening moves are known [6,12].

Typical Hex solvers use tree search and these optimizations [1] : a) search promising moves first (which saves time when the move wins), e.g.. with a neural network [4], b) prune provably redundant moves (which limits the growth of the tree size), c) detect wins early, e.g.. by computing virtual connections [8,13], d) use a transposition table to recognize positions already solved.

© Springer Nature Switzerland AG 2020
T. Cazenave et al. (Eds.): ACG 2019, LNCS 12516, pp. 79–90, 2020.
https://doi.org/10.1007/978-3-030-65883-0_7

In this paper we present new fill and prune results. This is of interest from a combinatorial game theory (CGT) point of view and can speed up solving and playing. Fill replaces a position (whole-board pattern) with an equivalent one with fewer empty cells, which can help all optimizations, especially (b). Section 2 gives definitions and results, Sect. 3 recalls known fill results, Sect. 4 presents new fill results, Sect. 5 presents new prune results, and Sect. 6 reports on experiments.

2   General Analysis

$s(X)$ denotes a stone (or move) of player X in cell s. E.g., position C changes to $C + s(W)$ after White plays in s. $\overline{X}$ denotes the opponent of player X. The two players are usually B and W. The following definition – which we will apply to positions that differ by only a few stones – relies on the relation $\leq$ from CGT. We will only use it with positions that differ by just a few stones.

Definition 1 (A pre-order relation). *For two positions C, C', we say that $C \leq_X C'$ when for any player Y, if player X has a winning strategy in C_Y to play, then she also has one in C'_Y to play.*
For two patterns P, P', we say that $P \leq_X P'$ when, for every position C that contains P, if we call C' the position where it is replaced by P', then $C \leq_X C'$.

Note 2. Hex is a perfect-information deterministic no-draw 2-player game, so $\leq_W$ is exactly $\geq_B$, so $=_W$ is the same as $=_B$, which we write as $=$.

As with Defenition 1, the following definitions and properties can be extended from positions to patterns: a pattern P has property A if, for every position C that contains P, C has property A.

Definition 3 (Dead cells and stones). *An empty cell s is* live *for player X when it is in a minimal (for inclusion) set of empty cells that, if X-colored, would join X's two sides. A cell is live for one player whenever it is live for the other [5]. A cell that is live for neither player is* dead. *A stone is* dead *when, if removed, the associated cell is dead.*

Theorem 4. *[5] If the empty cell s is dead, then $C = C + s(B) = C + s(W)$.*

In practice, dead cells can be detected either by seeing the board as a graph or by using patterns as those in Fig. 2.

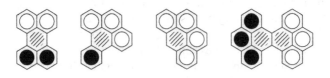

Fig. 2. Each dashed cell is dead. Any whitespace cell can be empty or colored.

Definition 5 (Fill pattern). *An empty cell s is X-fillable when* $C = C + s(X)$.

E.g., any dead cell is W-fillable and B-fillable. In practice, X-fillable cells are usually X-filled and then considered as stones.

Definition 6 (Inferior cells). *An empty cell s is X-inferior to another empty cell t when* $C + s(X) \leq_X C + t(X)$.

Definition 7 (Reversing a move). *A move* $r(X)$ *X-reverses* $s(\overline{X})$ *when* $C + r(X) + s(\overline{X}) \geq_X C$.

This CGT notion is useful: after $s(\overline{X})$, X should reply with $r(X)$. (Warning: if $s(\overline{X})$ is a blunder, X might have a better reply than $r(X)$.) In some (but not all) reverse patterns, $s(\overline{X})$ can be pruned, e.g.. with vulnerable patterns [5]. In §5.1 we give a condition sufficient to prune some reversible moves.

Definition 8 (X-iterativity). *A cell (resp. empty cell) s is X-iteratively dead (Y-filled) when we can iteratively X-fill some cells that leave s dead (Y-filled).*

Theorem 9. *If* $s(\overline{X})$ *is X-iteratively dead, then* $C = C - s(\overline{X}) = C - s(\overline{X}) + s(X)$.

Proof. Let T be the set of cells that is X-filled to kill s. Then, $C = C + T(X) = C - s(\overline{X}) + s(X) + T(X) \geq_X C - s(\overline{X}) + s(X) \geq_X C - s(\overline{X}) \geq_X C$.   □

Theorem 10. *If s is X-iteratively X-filled (and so empty) then* $C = C + s(X)$.

Proof. Let T be the set of cells that is X-filled to fill s. Then, $C = C + T(X) = C + s(X) + T(X) \geq_X C + s(X) \geq_X C$.   □

Some following results assume $C = C + s(X)$ as an hypothesis, so we can use iterative death/fill as sufficient conditions. This is also useful for pruning, in two ways:

– $C + s(X) = C + s(X) + t(X)$ implies that t is inferior to s for X, so provided that we explore s, or a move superior to s, we can prune t;
– $C + r(\overline{X}) + s(X) = C + r(\overline{X})$ implies that $s(X)$ is $\overline{X}$-reversible by $r(\overline{X})$.

3   Previous Fill Results

Dead and iteratively dead cells can be filled. Other fill results are known [7,10]:

Definition 11 (X-captured cells). *A set of empty cells S is X-captured when X has a second-player strategy in S that guarantees every cell in S to be X-colored or dead.*

Theorem 12. *If S is X-captured, then any subset of S can be X-filled.*

Proof. $C + S(X) \geq_X C$, and the strategy gives $C + S(X) \leq_X C$.   □

Fig. 3. Each white-dotted cell is W-captured.

In practice, captured cells are detected using patterns such as those in Fig. 3.

Definition 13 (X-permanently inferior cells). *Assume that there is an empty cell s and a set of empty cells T (with $s \notin T$) such that $\forall t \in T, P + t(\overline{X}) + s(X) = P + s(X) = P + t(X) + s(X)$. Then any cell $t \in T$ such that $P + t(X) + s(\overline{X}) = P + s(\overline{X})$ is said X-permanently inferior.*

Theorem 14. *If t is X-permanently inferior, then t can be X-filled.*

Proof. Let C be such a position and let $C' = C + t(X)$. In C (and in C'), each cell of T is X-inferior to s and $\overline{X}$-reversible by s, so we may assume that one player will play in s while T is still empty. But after this first move, regardless of who makes it, t can be filled. $\qquad\square$

Before, only the three patterns shown in Fig. 4 were known.

Fig. 4. Each white-dotted cell is W-permanently inferior.

4   New Fill Results

4.1   Near-Death Patterns

Definition 15 (Near-death pattern). *Pattern P with non-intersecting empty cell sets b and w and remaining empty cell set x is near-death when these hold:*

- *if b is B-colored and w is W-colored, then x is dead;*
- *if w is W-colored, then B has a second-player strategy in $b \cup x$ that leaves all cells of b black or dead;*
- *if b is B-colored, then W has a second-player strategy in $w \cup x$ that leaves all cells of w white or dead.*

We calls the cells of x nearly dead because, assuming optimal play, each player can save her side of the pattern and what remains will be dead:

Theorem 16. *For a near-death pattern P, $P = P + b(B) = P + w(W) = P + b(B) + w(W)$.*

Proof. First, we have $P + w(W) \geq_W P + b(B) + w(W) \geq_W P + b(B)$ and $P + w(W) \geq_W P \geq_W P + b(B)$, so it is sufficient to prove $P + b(B) \geq_W P + b(B) + w(W) \geq_W P + w(W)$, which by symmetry reduces to $P + b(B) \geq_W P + b(B) + w(W)$. Now following the white second-player strategy gives $P + b(B) \geq_W P + w(W) + b(B) + x(B) = P + w(W) + b(B)$.   □

Note 17. In the hypothesis, one might be tempted to replace $b \cup x$ by $w \cup b \cup x$ (and to remove "if w is W-colored"). This is sufficient for $P = P + b(B) + w(W)$ but not $P = P + w(W)$, so monocolor-iterativity cannot be used for pruning.

Note 18. When $b = \emptyset$ the theorem implies that W-captured cells can be W-filled, which is not new. But some capture patterns found with $b = \emptyset$ are new.

Here is how to produce patterns from this theorem: choose a core death pattern; remove some stones; add a gadget (pattern of stones and empty cells) so that, for each cell that had a stone, that cell can be killed by playing in an empty cell; check whether the theorem applies. See Fig. 5, 6 and 7.

Fig. 5. Start with a pattern where the dashed cell is dead. Remove one stone for each player. On the final pattern, where two gadgets have been added, apply the theorem with b containing the black-dotted cell, w containing the white-dotted cell and x containing the dashed cell.

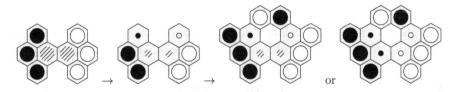

Fig. 6. A larger near-death pattern, constructed similarly. Here the theorem still applies if we move the left cell from x to b and the right one from x to w, giving a stronger result. This phenomenon often occurs when x has at least 2 cells.

This method gives more than 100 new fill patterns of radius 3 (each pattern has up to $1 + 6 + 12 + 18 = 37$ cells).

Fig. 7. A new W-capture pattern, constructed similarly

4.2 Mutual-Fill Patterns

Theorem 19. *Let P be a pattern with empty cells S, and two disjoint subsets T, U of S. Denote $P' = P + T(B) + U(W)$ and $S' = S \backslash (T \cup U)$. Assume this:*

$$\exists b \in S, P + b(B) = P + S(B); \qquad \exists w \in S, P + w(W) = P + S(W);$$
$$\exists b' \in S', P' + b'(B) = P + S(B); \quad \exists w' \in S', P' + w'(W) = P + S(W).$$

Then $P = P'$.

To prove the theorem, it suffices to use that in both P and P' whoever plays first in the pattern gets all of S. We omit the details.

Figure 8 shows the four mutual-fill patterns we found. The leftmost pattern applies on any empty board with at least two rows and columns.

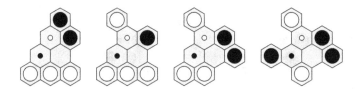

Fig. 8. In each pattern the dotted cells can be mutually filled.

Note 20. Warning: mutual-fill can change a move from losing to winning. E.g., apply the left-most pattern in Fig. 8 to the 2×2 board: now each corner cell wins, whereas on the original empty board it is losing.

Note 21. This theorem fills cells for both players at the same time, so it cannot be used to derive strong-reversible or inferior cells via monocolor-iterativity.

4.3 New X-permanently Inferior Patterns

Along with the permanently inferior patterns of Fig. 4, we found the three new ones shown in Fig. 9. The leftmost occurs often on a border.

We also found a fourth more complicated pattern, shown in Fig. 10, which can occur near an acute corner of the board. We omit the proofs that these patterns are permanently inferior. Combining the left patterns of Figs. 10 and 8 yields the right pattern of Fig. 10.

Fig. 9. Three new W-permanently inferior patterns.

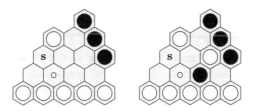

Fig. 10. Another new W-permanently inferior pattern, and a corollary.

5 New Prune Results

Previous prune results were based on mustplay (prune moves that let the opponent a simple way of making a winning connection) and monocolor-iterativity. This second approach is strengthened by the large number of new fill patterns. Other results were based on star decomposition [9] – some of which now apply via mutual-fill or near-death patterns – and on strategy-stealing arguments [1].

Prune results often have the form "if s wins, then t wins". Thus, we must take care to avoid logical implication cycles when pruning moves. Warning: some of our results here are based on game history, which requires extra consideration when implemented with a transposition table.

5.1 New Theorem Using Reversibility

Here we strengthen a result from [9].

Definition 22 (X-strong-reversing). *A move $r(X)$ X-strong-reverses $s(\overline{X})$ when $C + r(X) + s(\overline{X}) = C + r(X)$.*

Definition 23 (Carrier of a X-strong-reverse pattern). *For a X-strong-reverse pattern P where r X-reverses a single cell s, we denote by F the set of the empty cells except r and s. This is the* carrier *of the pattern.*

Definition 24 (Blocking). *A X-strong-reverse pattern P_1* blocks *another one P_2 when r_1 is in F_2.*

In particular, some patterns (usually called "vulnerable") have an empty carrier, so they are never blocked.

Fig. 11. (left) The black-dotted cell B-strong-reverses the white-dotted cell (carrier size 2). (right) The black-dotted cell B-strong-reverses each white-dotted cell (carrier size 1 each), and neither pattern blocks the other.

Theorem 25. *Let (P_n) be a sequence of (possibly overlapping) X-strong-reverse patterns of a position C. Assume that for all $k < n$, P_n does not block P_k, and that $\overline{X}$ has a winning move in C. Then $\overline{X}$ has a winning move that is not one of the s_n (or, if there is no other move, then skipping the next move wins).*

Proof. The special case where there is no other move is left as an exercise for the reader.

Argue by contradiction: assume that the theorem is false, and let $(P_n)_{1 \leq n \leq N}$ be a counter-example with N minimum, i.e. the only winning moves (with $\overline{X}$ to play) are among the s_n (so $N \geq 1$)

By applying the result to $(P_n)_{1 \leq n \leq N-1}$, we know that s_N wins, so $C + r_N(X) + s_N(\overline{X}) = C + r_N(X)$ (with $\overline{X}$ to play) wins for $\overline{X}$. In this new position, we can apply the result to (P_n), from which we remove the potential patterns for which $r_N = s_n$ (because $\overline{X}$ cannot play in s_n) or $r_N = r_n$ (but anyway in this case s_n is $\overline{X}$-fillable, so if s_n wins for $\overline{X}$ then any move or even skipping the next move wins) (this removes at least $n = N$), and we get that there is a winning move t that is not one of the s_n. This means that $C + r_N(X) + t(\overline{X})$ (with X to play) wins for $\overline{X}$, so t was a winning move for $\overline{X}$ in C, contradiction. □

5.2   New Reverse Patterns

Theorem 26. *Let P be a pattern with exactly 3 empty cells s, t, u such that, in $P + t(X) + u(X)$, s is dead, and $P + s(\overline{X}) + t(X) + u(\overline{X}) \geq_X P + s(\overline{X}) + t(\overline{X}) + u(X)$. Then $t(X)$ reverses $s(\overline{X})$.*

Proof. First, we prove that $P + s(\overline{X}) + t(X) \geq_X P + s(\overline{X}) + u(X)$. We denote $P_1 = P + s(\overline{X}) + u(X)$ and $P_2 = P + s(\overline{X}) + t(X)$. Let C_1 be a position containing P_1, and let C_2 be the position obtained from C_1 by replacing P_1 with P_2. Assume that C_1 wins for X with a strategy S. We will show that C_2 also wins for X.

To win in C_2, X follows S, pretending that there is P_1 instead of P_2 until one player moves in P_2. There are three cases:

- If neither player moves in P_2, then at the end of the game, after a X move, X would have a winning connection in C_1' and thus in $C_1' + t(\overline{X})$; and $C_1' + t(\overline{X}) \leq_X C_2' + u(\overline{X}) \leq_X C_2'$.

- If S tells X to play in t then X plays instead in u, and we get a position that S proves wins.
- If $\overline{X}$ plays in u, then S tells us that $C_1' + t(\overline{X})$ wins for X, and given $C_1' + t(\overline{X}) \leq_X C_2' + u(\overline{X})$ we get a position at least as good as one that S proves wins.

Now we prove that $P \leq_X P_2$. Let C be a position containing P, and let C_2 be the position obtained from C by replacing P with P_2. Assume that C wins for X with a strategy S. We will show that C_2 also wins for X.

To win in C_2, X follows S, pretending that there is P instead of P_2, until one player moves in P_2. Again, there are three cases:

- If neither player moves in P_2, argue as in the same case above.
- If S tells X to play somewhere in P, then X plays in u instead. Then s is dead so we can consider s to be X-colored, so we get a position at least as good as one that S proves wins.
- If $\overline{X}$ plays in P_2, then this must be in u given this is the only empty cell. Then X pretends that $\overline{X}$ has played in s in P, and follows S accordingly. P_2 is now full, so $\overline{X}$ can not play there any more. The case where X also does not play there can be treated as previously. So we may assume that at some point S tells X to play in P, and we may assume that this is in t since $P + s(\overline{X}) + t(X) \geq_X P + s(\overline{X}) + u(X)$. But then S tells us that, after this move and $\overline{X}$'s reply in u, X wins. So the current position wins.

$\square$

To produce patterns from this theorem, we combine one left-gadget (to make sure that W-coloring t and u kills s) with one right-gadget (to make sure that B-coloring s and t kills u, which in practice is the only interesting case) from Fig. 12. This yields 18 patterns, two of which are shown in Fig. 13.

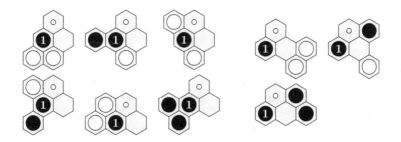

Fig. 12. The 1 (resp. dotted) (resp. empty) cell coresponds to s (t) (u). (left) The six left-gadgets. (right) The three right-gadgets.

Note 27. Unlike strong-reverse moves, which are mostly pruned (see Thm 25), a reversible move be good, or even a unique winning move. Often, when $s(\overline{X})$ is really good, $\overline{X}$ should move in $u(\overline{X})$ immediately after $t(X)$. We have not formalized this observation (in particular the definition of "really good").

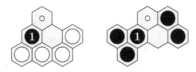

Fig. 13. In each pattern, the white-dotted cell W-reverses 1.

5.3   Self-reversibility

Observation 28. *Let s,t,u be empty cells such that* $C + s(X) + t(\overline{X}) + u(X) = C + t(\overline{X}) + u(X)$. *Then* $C + u(X) \geq_X C + s(X) + t(\overline{X}) + u(X)$.

In practice, this means that the move $u(X)$ after the moves $s(X), t(\overline{X})$ can be pruned (up to logical implication cycles), because X should have played in u instead of s. E.g., this can happen in the games in Fig. 14.

Fig. 14. In each game, after moves 1 and 2, White should not play in the white-dotted cell.

The conditions of this theorem are easily detected: for a possible move, can you prove (e.g.. by monocolor-iterativity) that it would make your previous move useless? If yes, then you should move elsewhere.

6   Experiments

We ran some experiments using the Benzene software package (maintained by Chao Gao at https://github.com/cgao3/benzene-vanilla-cmake) that includes the player MoHex and a parallel depth-first proof number (DFPN) search Hex solver [11,12].

We implemented near-death and new permanently inferior patterns. These are simply fill patterns, so this was straightforward. We did not implement mutual fill, due to technical issues mentioned in Note 20.

We also implemented applications of Theorem 25(in greedy fashion) and the new reversible patterns. We did not implement self-reversibility: we suspect that the cost of checking for logical implication cycles would yield negligible improvement.

We compared different implementations by solving a fixed set of 1-move Black openings on the 8 × 8 board. The results are shown in Fig. 15.

	none	Nd	Pi	F	F+St	F+St+Rp	F+St+Rp'	F+Rp'
a1	103432	93774	96544	93416	93398	95287	88844	88838
b2	341130	356745	316543	320501	327763	384242	316390	316695
c3	195830	193217	190741	195419	195423	240345	202943	202936
Σ	640392	643736	603828	609336	616584	719874	608177	608469
d4[1]	155659	200313	372527	198621	198649	184876	186944	361603
Σ	796051	844049	976355	807957	815233	904750	795121	970072

Fig. 15. Number of nodes created, depending on features enabled. Openings a1 and b2 lose, while c3 and d4 win.   Nd: near-death patterns and derived strong-reversible and inferior patterns; Pi: new permanently inferior patterns and derived patterns; F: Nd+Pi; St: strong-reversible theorem, replacing theorem from [9]; Rp: reversible patterns, reverser always picked; Rp': reversible patterns where, for OR (resp. AND) node n, reverser r is picked only when r's disproof (proof) number is less than or equal to the minimum proof (disproof) number among n's children.(For opening d4, after White's strongest reply d5, Black has two main winning moves: g4 and c5. The latter has a smaller search tree, with ~ 65000 nodes instead of ~ 240000. This explains the differences for this opening, depending on the initial choices. One cannot fairly compare features that differ on this g4-c5 choice.)

Our experiments show that the benefit of the new patterns is in most cases positive. Using St is less beneficial. These features did not help as consistently as we expected, perhaps in part because a small change in fill or prune can cause DFPN search to move to a different part of the tree.

Feature Rp' is always better than Rp, perhaps because the presence of a reverser is positively correlated with the presence of a move more easily proved to be winning. Feature Rp' appears globally positive.

Overall, our preferred setting is to enable all new features.

Note 29. Each new pattern requires more pattern matching and so slows node creation. There are few reversible patterns, so they slow runtime negligibly. But the new fill patterns yield thousands of inferior and strong-reversible patterns. Thus, the previous tests were run with only with patterns as large as in Fig. 5. E.g., we excluded the pattern in Fig. 6. The slowdown almost exactly compensates for the reduction in size of the resulting search tree, but – given that the slowdown is a constant factor while pruning can yield exponentially smaller trees – we expect that for larger boards using the new patterns will be beneficial.

Acknowledgements. We thank Chao Gao for many helpful comments and suggestions.

References

1. Arneson, B., Hayward, R.B., Henderson, P.: Solving hex: beyond humans. In: van den Herik, H.J., Iida, H., Plaat, A. (eds.) CG 2010. LNCS, vol. 6515, pp. 1–10. Springer, Heidelberg (2011). https://doi.org/10.1007/978-3-642-17928-0_1

2. Beck, A., Bleicher, M.N., Crowe, D.W.: Excursions into Mathematics. Worth, New York (1969)
3. Gale, D.: Game of hex and the brouwer fixed point theorem. Am. Math. Monthly **86**(10), 818–827 (1979)
4. Gao, C., Müller, M., Hayward, R.: Focused depth-first proof number search using convolutional neural networks for the game of hex. In: Proceedings of the Twenty-Sixth International Joint Conference on Artificial Intelligence (IJCAI-17), pp. 3668–3674 (2017)
5. Hayward, R., van Rijswijck, J.: Hex and combinatorics. Discrete Math. **306**(19–20), 2515–2528 (2006)
6. Hayward, R.B., Toft, B.: Hex, the full story. CRC Press, London (2019)
7. Henderson, P.: Playing and Solving Hex. PhD thesis, University of Alberta, Edmonton, Alberta, Canada, Fall (2010). https://webdocs.cs.ualberta.ca/~hayward/theses/ph.pdf
8. Henderson, P., Arneson, B., Hayward, R.B.: Hex, Braids, the Crossing Rule, and XH-Search. In: van den Herik, H.J., Spronck, P. (eds.) ACG 2009. LNCS, vol. 6048, pp. 88–98. Springer, Heidelberg (2010). https://doi.org/10.1007/978-3-642-12993-3_9
9. Henderson, P., Hayward, R.B.: Captured-reversible moves and star-decomposition domination in Hex. Integers **13**(CG1), 1–15 (2013)
10. Henderson, P., Hayward, R.B.: A Handicap Strategy for Hex, pp. 129–136. MSRI and Cambridge University Press (2015)
11. Huang, S.-C., Arneson, B., Hayward, R.B., Müller, M., Pawlewicz, J.: MoHex 2.0: a pattern-based MCTS hex player. In: van den Herik, H.J., Iida, H., Plaat, A. (eds.) CG 2013. LNCS, vol. 8427, pp. 60–71. Springer, Cham (2014). https://doi.org/10.1007/978-3-319-09165-5_6
12. Pawlewicz, J., Hayward, R.B.: Scalable parallel DFPN search. In: van den Herik, H.J., Iida, H., Plaat, A. (eds.) CG 2013. LNCS, vol. 8427, pp. 138–150. Springer, Cham (2014). https://doi.org/10.1007/978-3-319-09165-5_12
13. Pawlewicz, J., Hayward, R.B., Henderson, P., Arneson, B.: Stronger virtual connections in hex. IEEE Tran. Comput. Intel. AI Games **7**(2), 156–166 (2015)
14. Yamasaki, Y.: Theory of division games. Pub. Res. Inst. Math. Sci. **14**, 337–358 (1978)

Solving Cram Using Combinatorial Game Theory

Jos W. H. M. Uiterwijk[⊠]

Department of Data Science and Knowledge Engineering (DKE),
Maastricht University, Maastricht, The Netherlands
uiterwijk@maastrichtuniversity.nl

Abstract. In this paper we investigate the board game Cram, which is an impartial combinatorial game, using an $\alpha\beta$ solver. Since Cram is a notoriously hard game in the sense that it is difficult to obtain reliable and useful domain knowledge to use in the search process, we decided to rely on knowledge obtained from Combinatorial Game Theory (CGT).

The first and most effective addition to our solver is to incorporate endgame databases pre-filled with CGT values (nimbers) for all positions fitting on boards with at most 30 squares. This together with two efficient move-ordering heuristics aiming at early splitting positions into fragments fitting in the available databases gives a large improvement of solving power. Next we define five more heuristics based on CGT that can be used to further reduce the sizes of the search trees considerably.

In the final version of our program we were able to solve all odd × odd Cram boards for which results were available from the literature (even × even and odd × even boards are trivially solved). Investigating new boards led to solving two boards not solved before, namely the 3 × 21 board, a first-player win, and the 5 × 11 board, a second-player win.

1   Introduction

Cram is a variant of Domineering, which is a two-player perfect-information game invented by Göran Andersson around 1973. Both games were popularized to the general public in an article by Martin Gardner [7] (where Domineering was called CrossCram). Both games can be played on any subset of a square lattice, though mostly they are restricted to rectangular $m \times n$ boards, where m denotes the number of rows and n the number of columns.

Play consists of the two players alternately placing a 1×2 tile (domino) on the board. For Domineering the first player may place the tile only in a vertical alignment, the second player only horizontally; for Cram there is no such restriction: each player may place a tile vertically or horizontally. The first player being unable to move loses the game, his opponent (who made the last move) being declared the winner. Since the board is gradually filled, i.e., Domineering and Cram are converging games, the games always end, and ties are impossible. With these rules the games belong to the category of *combinatorial games*, for

© Springer Nature Switzerland AG 2020
T. Cazenave et al. (Eds.): ACG 2019, LNCS 12516, pp. 91–105, 2020.
https://doi.org/10.1007/978-3-030-65883-0_8

which a whole theory, the Combinatorial Game Theory (further CGT in short) has been developed [1,3,6].

The only difference between Domineering and Cram thus is that in the latter both players are allowed to place a domino both horizontally and vertically, i.e., in any position both players have the same possible moves. By this fact Cram belongs to the category of *impartial games* [1,3,6].

Cram has not received as much attention as Domineering in scientific research. $1 \times n$ Cram, also called Linear Cram, was proposed as early as 1934 by Dawson and completely solved by Guy and Smith in 1956 [10]. The sequence of values of the game (which is called ·**07** by them) is quite irregular, but turns out to be periodic from $n = 53$ onwards, with a period of 34. Cram was further described in several sources on CGT [1,3,6], where many values were given, but mainly for small boards. For the $2 \times n$ boards it was stated [3] that all even-width boards have value *0 and all odd-width boards value *1, though no formal proof was given. The first systematic investigation of larger Cram boards was the 2009 master thesis by Martin Schneider [15]. It reported the solution of $3 \times n$ boards up to $n = 9$, and the 4×5, 5×5 and 5×7 boards. More recently Lemoine and Viennot [12] extended this to $3 \times n$ boards up to $n = 18$, $4 \times n$ boards up to $n = 9$, and $5 \times n$ boards up to $n = 8$. A few later results (3×19, 3×20, 5×9, 6×7, and 7×7) were published on-line [13].

In a previous publication [17] we reported how we have constructed CGT endgame databases for Cram for all board sizes up to 30 squares. We also provided a formal proof that $2 \times n$ boards have values *0 and *1 for even and odd n, respectively. It was further shown in a preliminary experiment that incorporating most of these databases in a simple $\alpha\beta$ solver considerably raised the solving efficiency [20]. We now have developed an elaborated $\alpha\beta$ solver for Cram aiming at solving large Cram games more efficiently and especially solving even larger games not solved hitherto. Below we report the results.

2   Combinatorial Game Theory for Cram

For combinatorial games a whole theory has recently been developed, the Combinatorial Game Theory [1,3,6]. According to this theory the main distinction within combinatorial games is in *partisan games* and *impartial games*. In partisan games both players have their own moves, for which the types and CGT values may vary wildly. Domineering belongs to this category, and a survey on different (types of) values occurring in Domineering positions was given in [18]. In CGT a game position is defined by the sets of options (moves) for the two players, conventionally named Left and Right, separately. So the notation is

$$G = \{G_1^L, G_2^L, ... | G_1^R, G_2^R, ...\},$$

where $G_1^L, G_2^L, ...$ are the games (positions) that can be reached in one move by the left player, and similarly $G_1^R, G_2^R, ...$ for the right player. Options for Left and Right are defined similarly, thus full game definitions are stated in a recursive way and can become quite complex.

For partisan games the definitions are simpler: since both players have the same moves, a game position is just defined by the set of all its options

$$G = \{G_1, G_2, ...\}.$$

If options have the same value, then all but one can be removed from the set. Moreover, for impartial games the Sprague-Grundy Theorem [9,16] states that any position is equivalent to some Nim pile and the value of that position is then always a *nimber* $*n$, where n is the size of the equivalent Nim pile.

Note that in CGT it is common to use the terms "game position" and "game value" synonymously. This stems from a powerful theorem stating that game positions with the same value act the same and are fully interchangeable (even with positions or configurations of completely different games), so are *equivalent*. Therefore, if some game position has some value x, the game position is often identified by its value and we speak just about the game x.

2.1 Nimbers

Nimbers are a special type of CGT values occuring in impartial games and sometimes in partisan games. Their main characteristic is that the options follow a certain pattern. Before we give this pattern, we first introduce some small nimbers. By reason of self-containedness we repeat a short introduction to the theory of nimbers taken from [17].

The most common nimbers occurring in Cram are the endgame $*0$ and the star game $*1$. In game notation the endgame looks as $G = \{\}$. Its value is denoted by $*0$. It denotes a terminal position where neither player has a possible move. Therefore the endgame is a loss for the player to move. Due to the equivalence of game positions with equal values it means that any non-terminal Cram position with value $*0$ is also a loss for the player to move. Besides the trivial position with no squares at all, the position consisting of a single empty square (where no domino can be placed in either direction) is the only (but frequent) terminal position with value $*0$, see Fig. 1.

$$\square = \left\{ \ \right\} = *0$$

Fig. 1. A Cram endgame position.

A star is a position with just a single move, or equivalently, where each move leads to a terminal position $*0$. It looks as $G = \{*0\}$ and it is denoted by $*1$. It is of course a win for the player to move. Figure 2 shows several $*1$ positions in Cram.

Since the values of all these positions are equal, the positions are equivalent, meaning that they are interchangeable without changing the outcome of the game. The next nimbers, $*2$ and $*3$, are defined as $*2 = \{*0, *1\}$ and $*3 =$

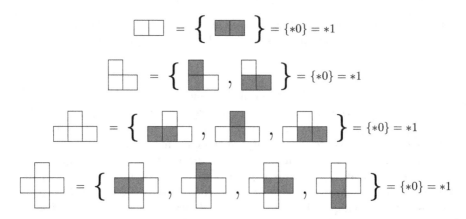

Fig. 2. Several equivalent ∗1 positions in Cram.

{∗0, ∗1, ∗2}, respectively. Example ∗2 and ∗3 positions are shown in Fig. 3, where the value of the third option of the ∗3 position is ∗0, since the player to move always will lose. Note that in this and following positions symmetric options (with same values as previous options) are omitted.

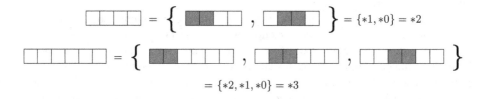

Fig. 3. Example ∗2 and ∗3 positions.

In general a nimber ∗n is defined as ∗n = {∗0, ∗1, ..., ∗(n − 1)}. It might be tempting to think that an empty 1 × 2n board has value ∗n, but that does not hold for n > 3. In fact the sequence of values, even for simple strips of 1 × n is quite chaotic and they are difficult to obtain, except by laboriously investigating all options (recursively). Note that it is not necessarily the case that all options of a position have all values from a consecutive series from ∗0 to ∗(n−1), to yield a ∗n position. More precisely, the value of an arbitrary Cram position is obtained as the *mex* function applied to the values of the options of the position. The *mex* function (minimal excludant) is defined as the smallest nimber not among the options. E.g., if a position has options with values ∗0, ∗1, ∗2, and ∗4, its value is ∗3.

As an example, consider the position depicted in Fig. 4. Since all options have value ∗0 or ∗2, the position has value ∗1.

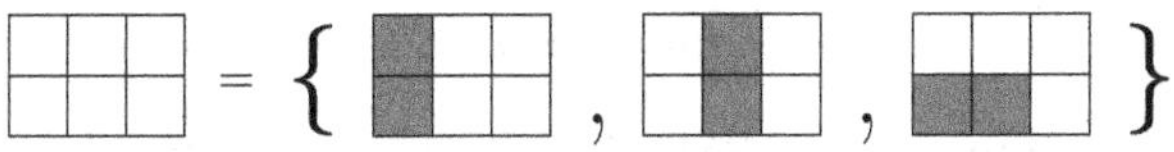

$$\boxed{\;\;\;\;} = \left\{ \boxed{\;\;} , \boxed{\;\;} , \boxed{\;\;} \right\} = \{*0, *0, *2\} = *1$$

Fig. 4. A $*1$ position with options with values $*0$ and $*2$ only.

This process of calculating the values of all options (and the options of the options, etc., up to endgame positions) is a laborious task, too time-consuming to execute during the search process of a solver. This is exactly why we opted for constructing Cram endgame databases filled with their nimber values once and for all.

What all nimbers $*n$ have in common is that if $n > 0$ the first player to move wins by moving to $*0$, and that if $n = *0$ the first player to move loses (all moves lead to positions with nimber value $> *0$, being wins for the next player).

2.2 Disjunctive Sums of Subgames

According to the Sprague-Grundy Theorem a Cram position with value $*n$ is equivalent to a single heap of n tokens in the game of Nim. Therefore, the theory for Nim can be applied to (sums of) Cram positions. As a consequence, if a position is decomposed into several disjoint subgames, the CGT value of the whole position is the Nim sum of the CGT values of the components. This Nim-addition rule was already discovered in 1902 by Bouton in analyzing the game of Nim [4]. In particular, if two Nim piles (or Cram fragments) have nimber values $*m$ and $*n$, their sum is given by the Nim sum $*m \oplus *n$, where $\oplus$ is the exclusive-or operator applied to the binary representations of m and n. As an example, in Fig. 5 a position is depicted with two non-interacting fragments with values $*1$ (binary representation 01) and $*2$ (binary representation 10). Since the Nim sum of 01 and 10 $= 11$, binary for 3, the position has value $*3$.

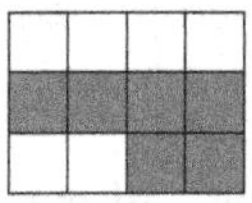

Fig. 5. A $*3$ position consisting of two fragments with values $*1$ and $*2$.

This also introduces an interesting property of nimbers: since the exclusive-or of any number with itself yields 0 ($n \oplus n = 0$) for any $n \geq 0$, every nimber is its own negative, i.e., $*n = -*n$. As a consequence, when two identical nimbers occur, they cancel each other, since their sum is equivalent to the endgame.

2.3 Cram Strategies

For empty rectangular positions in Cram it is easy to state some general observations regarding the CGT values. These were already given by Gardner [7].

Theorem 1. *Every even × even empty Cram board is a second-player win.*

Proof. The second-player's strategy to reach this is easy: after every first-player's move the second player responds by playing the symmetrical move (with respect to the centre of the board), ensuring that he/she will make the last move and win the game. Consequently, even × even empty boards have value *0.   □

Theorem 2. *Every even × odd empty Cram board is a first-player win.*

Proof. The winning strategy of the first player is to occupy with the first move the two central squares, after which he/she plays according to the strategy given in Theorem 1 (pretending to be the second player from now on), securing the win. Consequently, even × odd (and odd × even) empty boards have values $*n$ with $n \geq 1$.   □

Regarding odd × odd empty boards both theorems give no clue about their values, and no other optimal strategy is known, so they can be either first-player wins (with any positive nimber value) or second-player wins (with value *0).

3 Experiments and Discussion

We implemented a straightforward depth-first $\alpha\beta$ [11] solver in C# on a standard HP desktop PC with an Intel i5-4590 CPU @ 3.30 GHz. This searcher investigates lines until the end (returning that either the first or the second player made the last move and so wins the game). We further used a standard transposition table [8] based on Zobrist hashing [22] using a Deep replacement scheme [5]. This table contains 2^{26} entries of 8 bytes each, so with a total size of 0.5 GB. Further the Enhanced Transposition Cutoff (ETC) method [14] is implemented. Mirror symmetries and (for square boards) rotational symmetries are taken into account.

The CGT endgame databases used [17] consisted of all databases with a size up to 30 squares, with 1 byte per entry (representing a nimber). In total they need 3.77 GB of storage.

To make the experiments feasible, all experiments were terminated when a threshold of 1×10^9 nodes investigated is exceeded, unless noted otherwise.

3.1 Base Case

As a base case we first performed experiments with a plain $\alpha\beta$ solver without any particular heuristics[1] or CGT knowledge. We did experiments with several move-ordering heuristics, but these did not yield any significant increase in efficiency so far. This is in agreement with our experience that it is extremely difficult to foresee which moves on a board will be winning and which ones losing. Therefore we decided to run these experiments without any move ordering. This means that

---

[1] By a heuristic we mean a method that does not impact the final results, but only (hopefully) obtains the results faster.

in a position first all possible horizontal moves are investigated, from left to right and then down to up, and then all possible vertical moves, in the same order. Since according to Sect. 2.3 all even × even boards are losses for the first player, whereas all odd × even (and so even × odd) boards are wins for the first player, we only investigated odd × odd boards. Results using this basic $\alpha\beta$ solver are given in Table 1.

Table 1. Number of nodes investigated solving odd × odd Cram boards using several search engines (explained in the text). A '−' in an entry denotes that the board could not be solved within the threshold of 1×10^9 nodes investigated.

Board	win	BF	+ DB	+ FN	+ FS
3 × 3	2	7	1	1	1
3 × 5	1	254	1	1	1
3 × 7	1	7,617	1	1	1
3 × 9	1	415,550	1	1	1
3 × 11	2	14,801,591	413,522	192,004	31,109
3 × 13	1	−	16,061,481	3,032,649	80,917
3 × 15	1	−	733,255,885	22,351,558	73
3 × 17	2	−	−	−	627,228,056
5 × 5	2	73,500	1	1	1
5 × 7	1	49,687,795	4,496,711	4,906,301	1,978,017
5 × 9	1	−	−	−	142,237,539
7 × 7	1	−	−	−	956,663,573

In this table the first column gives the board dimensions, the second column the winner (a '1' for a first-player win and a '2' for a second-player win) and the third column (marked 'BF' for Brute Force) gives the number of nodes investigated to solve the board. The remaining columns denote the number of nodes investigated when some particular enhancement is added (discussed below). To keep the number of experiments feasible, we have chosen to always add a single new enhancement to the previous one.

3.2   Using Endgame Databases

Our next step is to investigate how the databases can increase the efficiency. For that we check during search whether the position consists of only a single or more fragment(s) available in the databases. If not, the search continues, otherwise if the position consists of a single fragment, the position is denoted as a win for the player to move (value $\geq *1$) or for the opponent (value $*0$), and the branch is pruned. If the position consists of more fragments the Nim sum of the fragments is calculated and used to value the position accordingly. The results using this database support are tabulated in Table 1 in the column marked '+ DB'.

It is clear that database support has a profound impact on the solving efficiency of our Cram solver. Of course all boards with a size of at most 30 squares just need one investigated node now, since the results are directly available from the databases. But also for all other solved boards we see a large increase in efficiency. Moreover, two boards that were not solvable within the threshold without database support are now solvable (3×13 and 3×15).

Once database support is incorporated, two move-ordering heuristics come to mind, both aiming at early splitting the position in multiple fragments.

Fragment Narrowing. The easiest of the two to implement is denoted as the *Fragment Narrowing Heuristic* (FN). According to this heuristic, for non-square boards if the board is "wide" (more columns than rows) vertical moves are always preferred over horizontal moves, and vice versa for "high" boards (more rows than columns). This is justified since moves across the short dimension will in general split the position earlier into subfragments. As a second criterion, both for vertical and horizontal moves, we prefer moves closer to the centre. The reason for this is that when a move splits the position into fragments, it is better that they are of roughly equal size than one small and one large fragment, since in the latter case the probability is higher that the larger fragment still will not be available in the databases. The results using this heuristic additionally to the previous version are tabulated in Table 1 in the column marked '+ FN'.

It is clear from the table that this heuristic also has a significant positive influence on the solving efficiency.

Fragment Splitting. As second move-ordering heuristic we make sure that moves splitting a fragment are preferred over moves just narrowing a fragment (since the narrowing has as goal an early splitting). This is implemented as the *Fragment Splitting Heuristic* (FS) and has preference (by scaling the values) over the Fragment Narrowing Heuristic. This heuristic works by summing the squared sizes of all empty fragments and deducting these from the squared total number of empty fields on the board. In this way splitting moves are favoured and especially splitting moves where the parts are roughly equal in size. A simple example calculation of a move-ordering value will clarify this heuristic[2]: in the position on the 3×19 board after 1. j2:j3 the non-splitting move 2. i2:i3 with 1 fragment of 53 squares receives an ordering value of $53^2 - 53^2 = 0$. In fact, any non-splitting move gets a value of 0. The splitting move 2. i1:i2 with 2 fragments of 25 and 28 squares receives the ordering value $53^2 - 25^2 - 28^2 = 1400$. The optimal splitting move is 2. i1:j1 with 2 fragments of 26 and 27 squares receiving the ordering value $53^2 - 26^2 - 27^2 = 1404$. The results using this heuristic additionally to the previous version are tabulated in Table 1 in the column marked '+ FS'.

---

[2] We use chess-like notation for moves, where squares are indicated by their column ('a', 'b', etc, from left to right) and their row ('1', '2', etc, from bottom to top). A move consists of the two squares covered separated by a colon.

Again we observe a large improvement in solving power. Moreover, another three boards that were not solvable within the threshold without the FS Heuristic are now solved (3×17, 5×9, and 7×7).

Remarkable is the increase in solving efficiency for the 3×15 board. After the opening move h1:h2 (chosen due to the FN Heuristic) a position is reached (see Fig. 6) where all possible moves by the second player are trivially shown to be losses using the databases for $3 \times n$ with $n \leq 8$.

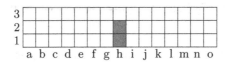

Fig. 6. The winning move h1:h2 on the 3×15 board.

Of the 66 possible moves, we only consider the 33 moves in the left half of the board (the other ones are symmetric). Using the FS Heuristic, first the moves that split the board are considered (g2:g3, g3:h3, and f3:g3). They lose immediately, since the Nim sum of the fragments after splitting is $\geq *1$. After all 30 other second moves, at level 3 first splitting moves are investigated, and for these boards it is always the case that one of them wins (leading to two fragments with Nim sum $*0$). As a consequence, the complete search tree for the 3×15 board is at most 3 deep, which leads to a tremendous reduction in tree size (from more than 22 million without the FS Heuristic to just 73 nodes with this heuristic).

3.3 Using CGT Values

In this subsection we investigate several ways to reduce the search trees further, based on using knowledge from CGT applied to Cram fragments obtained. The results are in Table 2. This table continues from the best results in Table 1 (the rightmost column marked '+ FS'), where we now omitted the results for the boards directly available from the databases. The remaining columns contain the experimental results for five enhancements explained below the table, again in an additive manner.

Skipping Moves. When the position is split in multiple fragments and one or more of these have known values (from the databases), but the value of at least one fragment is still unknown (so the search should continue), we often can omit moves in the known fragments from further consideration. We consider the following three steps:

1. In each fragment with value $*0$ all moves can be skipped from further investigation, since it denotes a losing fragment and it is never useful to play within

Table 2. Number of nodes investigated solving odd × odd Cram boards using several search engines (explained in the text). A '-' in an entry denotes that the board could not be solved within the threshold of 1×10^9 nodes investigated. In the final version (rightmost column) no threshold was set.

Board	Win	Table 1	+ SM	+ SF	+ LBS	+ SLR	+ BD
3 × 11	2	31,109	31,008	30,670	30,323	29,882	26,437
3 × 13	1	80,917	80,312	79,573	79,080	77,300	66,079
3 × 15	1	73	73	73	73	73	73
3 × 17	2	627,228,056	377,565,993	258,704,721	256,011,532	250,196,286	204,860,712
3 × 19	2	–	–	–	–	–	18,327,715,035
3 × 21	1	–	–	–	–	–	962,810,282
5 × 7	1	1,978,017	1,935,662	1,888,485	1,861,601	1,803,369	1,354,483
5 × 9	1	142,237,539	116,023,958	98,784,937	97,365,853	94,970,220	75,190,993
5 × 11	2	–	–	–	–	–	204,521,419,098
7 × 7	1	956,663,573	710,643,974	607,605,602	590,066,547	567,193,292	433,712,107

a losing fragment (every move leads to a new fragment with some value $*n$ with $n \geq 1$, to which the opponent always can just respond with a move leading to a fragment with value $*0$ again).

2. Next we check if in the remaining known fragments pairs of fragments occur with the same value. Each such pair can be omitted from further consideration, since an arbitrary move by the player to move in one fragment of such pair, changing its value to some new value, can always be responded by the opponent in playing an equivalent move (to the same new value) in the other fragment of the pair. Therefore all moves in each such pair can be skipped. This of course also follows from the fact that the Nim sum of such a pair has CGT value $*0$ ($*n + *n = *0$ for arbitrary n).

3. Finally, we calculate the sum-value of all remaining known fragments that are not skipped. If this sum-value is $*0$ the fragments together are losing and all moves in them can be skipped (e.g., consider three fragments with values $*1$, $*2$, and $*3$, with Nim sum $*0$). If the sum-value is not $*0$, but lower than or equal to the value of one of the non-skipped fragments, then all fragments except this one can be skipped, since any value that can be reached by playing in one of all non-skipped fragments, can also be reached by playing in that particular fragment.

These three ways of skipping moves based on CGT-values of fragments are taken together in the *Skip Moves Heuristic* (SM). The results using this heuristic additionally to the previous version are tabulated in Table 2 in the column marked '+ SM'.

Although the increase of efficiency is negligible for the smaller boards, it still is significant for larger boards.

Simplifying Fragments. For all remaining known (unskipped) fragments it is useful to simplify such fragments if possible. This is based on the idea that two different fragments with the same value are equivalent. So if the fragment under consideration is large but is equivalent with a much smaller fragment, we simplify it. Effectively this is done by only allowing some specific moves in the fragment. More explicity, if a fragment has value $*n$, then we only allow n moves in this fragment, one resulting in a value $*0$, one in $*1$, up to one in $*(n-1)$. All other moves in the fragment are skipped. This method is called the *Simplify Fragments Heuristic* (SF). The results using this heuristic additionally to the previous version are tabulated in Table 2 in the column marked '+ SF'.

We observe again that the increase of efficiency is negligible for the smaller boards, but significant for larger boards.

Linear and Bended Fragments. Occasionally we encountered during search a linear chain that is longer than the dimension of any of the databases used. Since Linear Cram is completely solved for arbitrary length [10], we just obtain its value by a look-up in a prefilled array. As an example, the linear chain in Fig. 7(left) is not in any of our databases, still its value is determined to be $*5$. Consequently the whole position in this figure has value $*5$ and thus the position is a win for the player to move. More often, we encounter bended chains (also called *snakes* [3,21]) that have the same value as a linear chain of the same size. So these also can be looked-up in our preprogrammed array. As an example, the snake in Fig. 7(right) has the same value $*5$ as the chain in Fig. 7(left). The recognition of linear or bended snakes is simple: the fragment must consist of squares where exactly two have one neighbour (the endpoints of the snake), and all others have exactly two neighbours.

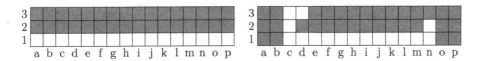

Fig. 7. An example of a linear chain of size 16 and a bended snake with the same size on the 3×16 board.

The method of recognizing and valuing snakes is called the *Linear/Bended Snakes Heuristic* (LBS). The results using this heuristic additionally to the previous version are tabulated in Table 2 in the column marked '+ LBS'.

Using this heuristic we observe a small increase in efficiency for all boards investigated.

Skipping Lost Responses. When in a position we investigate a move x for a player and if x turns out to be a loss after some response y by the opponent, then it is not needed for the player to investigate y itself (if not already done), since then y will lose to x now being played by the opponent. Of course this stems from the fact that the resulting position is the same (losing, so with value $*0$), and does not depend on the order of the moves played leading to the position. We denoted this as the *Skip Lost Responses Heuristic* (SLR). The results using this heuristic additionally to the previous version are tabulated in Table 2 in the column marked '+ SLR'.

Using this heuristic we observe a small but significant increase in efficiency for all boards investigated.

Bridge Destroying. In the last heuristic applied we use proofs in *Winning Ways* [3], showing that a Cram fragment with some particular pattern may be split into two fragments, such that the value of the original fragment (when not obtained from the databases) is equal to the sum of the values of the two smaller fragments (each with possibly known value). We distinguish four cases, according to the sketches depicted in Fig. 8.

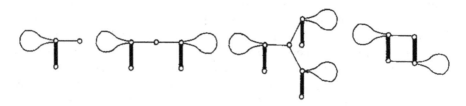

Fig. 8. Sketches of empty Cram fragments that can be broken in subfragments without changing the value. [Taken from [3], Vol. 3, p. 504.]

In these figures the dots denote empty squares, the lines denote connected (neighbouring) squares and the balloons denote arbitrary empty fragments (but not connected to the other parts of the fragment). Thin lines denote connections that may be broken without changing the value of the fragment.

The first 3 fragments have in common that there is a denoted empty square (the "bridge") connected to 1 to 3 subfragments, where each subfragment is connected to one single empty square (the bold connection) and further arbitrarily to some remainder (the balloon). Since in Cram a square has maximally four neighbours, this in principle leads to 4 such cases, but the fourth can "physically" only occur on boards where all balloons in the four subfragments are empty, thus consisting necessarily of a single empty square with four neighbouring fragments of exactly two empty squares each (a "cross"), with known value from the 5×5 database of $*0$. The first two (with one or two subfragments) occur frequently in Cram positions, the third one (with three subfragments) can physically only occur on boards with both dimensions ≥ 5 and moreover at least 1 of the three balloons empty, but still is useful.

The fourth fragment in Fig. 8 consists of four empty squares in a 2×2 arrangement with two arbitrary subfragments connected to two **opposing** squares of the 2×2 array. Note that in this case breaking the thin connections means that the fragment can be split either vertically through the 2×2 part (as depicted) or horizontally (rotated), but not diagonally. Again, the value of the whole fragment is equal to the sum of the values of the parts after splitting.

We implemented all four cases in Fig. 8 and denoted them together as the *Bridge Destroying Heuristic* (BD). The results using this heuristic additionally to the previous version are tabulated in Table 2 in the column marked '+ BD'.

Using this heuristic we observe a modest increase in efficiency for all boards investigated. We also solved a new board within the set threshold using this version of our solver, namely the 3×21 board. This board was hitherto not solved, so is our first extension of known solved Cram boards. It is a first-player win.

Since this is (for the time being) the last version of our Cram solver, we decided to also investigate a few more boards without using the 1×10^9 nodes threshold. First of all we investigated the 3×19 board (which remarkably needs much longer to be solved than the 3×21 board), requiring some 1.8×10^{10} nodes investigated. With considerably more effort we also solved the 5×11 board, needing some 2×10^{11} nodes. This board was also not solved up to date and is a second-player win.

4   Conclusions and Future Research

The results using the CGT endgame databases into our $\alpha\beta$ solver are good, with reductions of at least 90% in worst case for non-trivial boards, and of course 100% for the trivial boards (i.e., directly available from a database). This shows that the effect on solving power for the impartial game Cram is comparable to those for the partisan game Domineering [2] and the partisan all-small game Clobber [19]. By this result it is shown that this method of incorporating CGT into $\alpha\beta$ solvers by building CGT endgame databases is beneficial for a wide range of combinatorial games with the characteristic that game positions can split into independent subgames.

Using two move-ordering heuristics (the Fragment Narrowing Heuristic and the Fragment Splitting Heuristic) promoting early splitting of positions into independent fragments and thus enlarging the probability of early hitting the databases, results in another reduction in search tree size of a factor of 2 to 10 (and sometimes much more).[3] We can therefore conclude that endgame databases with CGT values are pivotal for the success of solving Cram boards.

The inclusion of five additional enhancements, based on domain-specific knowledge from Combinatorial Game Theory for Cram, enables a further large increase in solving power. Although modest per heuristic, in combination they reduce the search tree size for large boards with another factor of 2 to 3.

---

[3] The fact that these move-ordering heuristics are so successful does not contradict our statement that it is hard to find criteria indicating which moves are probably good or bad. It just leads to investigating first moves with small search trees (including hopefully "accidentally" winning moves).

All our results are in agreement with previous outcomes published in the literature. Moreover, we were so far able to solve two new Cram boards not solved before, the 3×21 board, being a first-player win, and the 5×11 board, being a second-player win.

Regarding future research a first idea is based on the observation that terminal nodes might vary widely regarding the depth in the search tree, even among splitting or narrowing moves. Therefore it might be useful to implement some form of iterative deepening into the search. We further still have several ideas based on Combinatorial Game Theory that might prune the search trees for Cram boards even further. Finally we consider implementing a Nimber Search program that calculates the exact CGT value for each position investigated instead of just determining if the position is a win or loss for the player to move. Comparing the results using Nimber Search with the results from our current $\alpha\beta$ solver it is interesting to investigate the claim by Lemoine and Viennot [12] that it might be more efficient to determine outcomes of Cram boards by computing nimbers for fragments than to determine the outcomes by developing directly the game tree for the sum of fragments.

References

1. Albert, M.H., Nowakowski, R.J., Wolfe, D.: Lessons in Play: An Introduction to Combinatorial Game Theory. A K Peters, Wellesley (2007)
2. Barton, M., Uiterwijk, J.W.H.M.: Combining combinatorial game theory with an α-β solver for Domineering. In: Grootjen, F., Otworowska, M., Kwisthout, J. (eds.) BNAIC 2014 - Proceedings of the 26th Benelux Conference on Artificial Intelligence, pp. 9–16. Radboud University, Nijmegen (2014)
3. Berlekamp, E.R., Conway, J.H., Guy, R.K.: Winning Ways for your Mathematical Plays. Academic Press, London (1982). 2nd edition, in four volumes: vol. 1 (2001), vols. 2, 3 (2003), vol. 4 (2004). A K Peters, Wellesley
4. Bouton, C.I.: Nim, a game with a complete mathematical theory. Ann. Math. **3**, 35–39 (1902)
5. Breuker, D.M., Uiterwijk, J.W.H.M., van den Herik, H.J.: Replacement schemes for transposition tables. ICCA J. **17**(4), 183–193 (1994)
6. Conway, J.H.: On Numbers and Games. Academic Press, London (1976)
7. Gardner, M.: Mathematical games. Sci. Am. **230**, 106–108 (1974)
8. Greenblatt, R.D., Eastlake, D.E., Crocker, S.D.: The Greenblatt chess program. In: Proceedings of the AFIPS Fall Joint Computer Conference, vol. 31, pp. 801–810 (1967)
9. Grundy, P.M.: Mathematics and games. Eureka **2**, 6–8 (1939)
10. Guy, R.K., Smith, C.A.B.: The G-values of various games. Proc. Camb. Philos. Soc. **52**(3), 514–526 (1956)
11. Knuth, D.E., Moore, R.W.: An analysis of alpha-beta pruning. Artif. Intell. **6**, 293–326 (1975)
12. Lemoine, J., Viennot, S.: Nimbers are inevitable. Theoret. Comput. Sci. **462**, 70–79 (2012)
13. Lemoine, J., Viennot, S.: Records. http://sprouts.tuxfamily.org/wiki/doku.php?id=records#cram. Accessed 13 May 2019

14. Plaat, A., Schaeffer, J., Pijls, W., de Bruin, A.: Nearly optimal minimax tree search? Technical report 94-19. University of Alberta, Department of Computing Science (1994)
15. Schneider, M.: Das Spiel Juvavum. Master thesis, Universität Salzburg (2009)
16. Sprague, R.P.: Über mathematische Kampfspiele. Tohoku Math. J. **41**, 438–444 (1935)
17. Uiterwijk, J.W.H.M.: Construction and investigation of Cram endgame databases. ICGA J. **40**(4), 425–437 (2018)
18. Uiterwijk, J.W.H.M., Barton, M.: New results for Domineering from combinatorial game theory endgame databases. Theoret. Comput. Sci. **592**, 72–86 (2015)
19. Uiterwijk, J.W.H.M., Griebel, J.: Combining combinatorial game theory with an α-β solver for Clobber: theory and experiments. In: Bosse, T., Bredeweg, B. (eds.) BNAIC 2016 – Artificial Intelligence. Communications in Computer and Information Science, vol. 765, pp. 78–92 (2017)
20. Uiterwijk, J.W.H.M., Kroes, L.: Combining combinatorial game theory with an alpha-beta solver for Cram. In: Atzmueller, M., Duivesteijn, W. (eds.) BNAIC 2018: 30th Benelux Conference on Artificial Intelligence, pp. 267–280. Jheronimus Academy of Data Science, 's-Hertogenbosch (2018)
21. Wolfe, D.: Snakes in Domineering games. Theoret. Comput. Sci. (Math Games) **119**, 323–329 (1993)
22. Zobrist, A.L.: A new hashing method with application for game playing. Technical report #88, Computer Science Department, The University of Wisconsin, Madison (1970). Reprinted in ICCA J. **13**(2), 69–73 (1990)

Exploiting Game Decompositions
in Monte Carlo Tree Search

Aline Hufschmitt[1]([✉]), Jean-Noël Vittaut[2], and Nicolas Jouandeau[3]

[1] CREC Saint-Cyr, Écoles de Coëtquidan, Guer, France
`aline.hufschmitt@st-cyr.terre-net.defense.gouv.fr`
[2] LIP6, CNRS, Sorbonne Université, F-75005 Paris, France
`jean-noel.vittaut@lip6.fr`
[3] University Paris 8, Vincennes-Saint-Denis, France
`n@up8.edu`

Abstract. In this paper, we propose a variation of the MCTS framework to perform a search in several trees to exploit game decompositions. Our Multiple Tree MCTS (MT-MCTS) approach builds simultaneously multiple MCTS trees corresponding to the different sub-games and allows, like MCTS algorithms, to evaluate moves while playing. We apply MT-MCTS on decomposed games in the *General Game Playing* framework. We present encouraging results on single player games showing that this approach is promising and opens new avenues for further research in the domain of decomposition exploitation. Complex compound games are solved from 2 times faster (*Incredible*) up to 25 times faster (*Nonogram*).

Keywords: Monte Carlo Tree Search · General Game Playing · Decomposition

1 Introduction

General Game Playing (GGP) is a branch of Artificial Intelligence with the aim of achieving versatile programs capable of playing any game without human intervention. Game specific algorithms cannot be used in a general game player as the game played should not be known in advance. An important aspect of GGP research is the development of automated rule analysis techniques to speedup the search.

Among the games considered in GGP, some are composed of different independent sub-games assembled sequentially or in parallel and played synchronously or asynchronously [3]. A player program able to identify the sub-games, solve them individually and synthesize the resulting solutions, can greatly reduce the search cost [2,4]. Some approaches have been proposed to decompose single [5] and multi-player games [6,7,17]. Using these decompositions two different strategies have been proposed to solve the global game in the GGP framework. The first approach, inspired from hierarchical planning and named *Concept Decomposition Search* [5], aims at solving single player games. The search algorithm is split into two stages: local search collecting all local plans

© Springer Nature Switzerland AG 2020
T. Cazenave et al. (Eds.): ACG 2019, LNCS 12516, pp. 106–118, 2020.
https://doi.org/10.1007/978-3-030-65883-0_9

and global search trying to interleave local plans from different sub-games to find the best global plan. These two steps are embedded into an iterative deepening framework. *Concept Decomposition Search* is extended to multi-player games [17] using *turn-move sequences* (TMSeqs) as a result of the local search. A TMSeq indicates both moves and players who performed them. Global search is based on standard search techniques but uses TMSeqs instead of legal moves, which results in a much smaller game tree compared to full search. The second approach [3] is based on Answer Set Programming (ASP) to solve decomposed games. Local plans are combined as and when they are calculated to find a global plan. This search is used on single player games. How to generalize this approach to multi-player games is not clear and remains an open problem. In these previous works, if a global plan is not found in the allocated time, the search returns nothing.

In this paper, we propose a new approach to solve decomposed games based on Monte Carlo Tree Search (MCTS). MCTS is the state of the art framework for GGP players but also for specialized player like Alpha Go [15], for video games and non-game domains [1]. Our Multiple Tree MCTS (MT-MCTS) approach builds simultaneously multiple MCTS trees corresponding to different sub-games. Instead of producing a global plan, MT-MCTS allows to evaluate moves in the different states of the game. A player program using MT-MCTS can therefore begin to play while continuing to explore the game and identify the next best moves. We compared the performance of our MT-MCTS algorithm with that of an MCTS search using the Upper Confidence Bound applied to Trees (UCT) policy and transposition tables on single player games.

The remainder of this paper is organized as follows. In the next section, we briefly review the MCTS framework, the UCT policy and common optimizations. In Sect. 3, we present our MT-MCTS approach using multiple trees to solve decomposed games. We present experimental results on several single-player games in Sect. 4. In Sect. 5, we discuss these results, the challenges risen by a simultaneous search in several trees, the possible extensions of our algorithm and open problems. We conclude in Sect. 6.

2   MCTS and UCT

The MCTS framework allows to build a game tree in an incremental and asymmetric manner [1]. Each node represents a state of the game, and each edge represents a joint move[1]. MCTS begins from a node representing the current state and then repeats four steps for a given number of iterations or for a given time: the selection of a path in the game tree according to some *tree policy*; the expansion of the tree with the creation of a new node; a game simulation (*playout*) according to a *default policy*; the back-propagation of the playout result to update nodes statistics. The number of playouts in MCTS is a key factor for the quality of the evaluation of a game tree node [9].

The *default policy* usually consists in playing random moves until a terminal state is reached. For the selection, UCT is the most common *tree policy*. It

---

[1] In the GGP framework, player moves are always simultaneous. In alternate move games, all players but one play a useless move to skip the turn.

provides a balance between the exploitation of the best moves and the exploration of the whole tree. A common optimization consists in using *transposition tables*: identical states are represented by a single node in the tree. In the GGP domain, many games that involve cycles use a *stepper*, a counter used to avoid infinite games. Identical states present at different depths in the game tree are then differentiated by this *stepper* and transpositions can only occur at a same depth. Using transpositions, a GGP game tree becomes a directed acyclic graph. When using transpositions, the evaluation of the different moves is commonly stored in the outgoing edges instead of the child nodes [14]. The number of visits remains stored in the parent node. Another common optimization used to guide the search is the pruning of fully explored branches [11,16]. During the selection step, instead of returning in branches that are completely evaluated, the mean score of the branch is computed and used for the back-propagation step.

3   Multiple Tree MCTS (MT-MCTS)

The decomposition of a game into several sub-games produces sub-states, in which available moves depend on the sub-states combination, and then provides a difficulty for legal moves computation. The decomposition also poses a problem for the identification of terminal sub-states. For example, the game *Incredible* is decomposed into a labyrinth (*Maze*), a game of cubes (*Blocks*), a *stepper* and a set of useless contemplation actions. The game is over if the *stepper* reaches 20 or if the *Maze* sub-game is over. But, in *Blocks*, a sub-state is never terminal by itself. We can also imagine a game where two sub-states are both non-terminal but their conjunction is terminal and must be avoided in a global plan. The decomposition also raises an issue for evaluating sub-states where scores can result from a timely combination with other sub-states. More specifically in GGP, the score described by the `goal` predicate is reliable only if the current state is terminal [10]. These two facts make the score function less reliable in sub-trees. At last, the decomposition raises the problem of its reliability. If the decomposition is inconsistent, the evaluation of legal moves can be wrong, leading the player program to choose illegal moves and compute inconsistent sub-states.

To avoid all these problems, we propose the following approach: doing simulations in the global game and building a sub-tree for each sub-game. Legal moves, next states and the end of game can be evaluated for the global game in which the real score is known. Move selection is performed according to the evaluation of the sub-states in the sub-trees. An inconsistency of the decomposition is detected if during two simulations, the same move from the same sub-state leads to different following sub-states. A partial[2] but consistent decomposition allows to play by the rules, although exploration may be less effective.

When a *stepper* is separated from the rest of a game, cycles can occur in some sub-games and in sub-state transpositions a move evaluation may differ according to the game depth. This problem is referred as the *graph history interaction problem* [12]. A general solution for games with binary scores is available [8].

---

[2] A decomposition is *partial* if a sub-game can be further decomposed.

However, in the GGP framework, the scores are more graduated and this general solution is therefore not applicable. To exploit some transpositions while avoiding the *graph history interaction problem*, the current version of our MT-MCTS considers transpositions only at a same depth in the sub-trees i.e. sub-games are represented by rooted directed acyclic graphs.

Global Simulations and Sub-trees Building: Our MT-MCTS iterates four steps like MCTS (Algorithm 1) except that the selection step is composed of alternated local and global selections[3].

Algorithm 1. MtMcts(nbPlayouts)

1: **for** nbPlayouts **do**
2:     S ← *current_state*
3:     $\{s_1, ..., s_n\}$ ← getSubstates(S)
4:     $\{val_1, ..., val_n\}$ ← selectionWithExpansion($\{s_1, ..., s_n\}$)
5:     simulationWithBackProp($\{val_1, ..., val_n\}$)

Global variables are:

S : the current global state
$\mathcal{M}$: moves played during selection and expansion
$\mathcal{S}$: sets of sub-states visited during selection and expansion
$\{val_1, ..., val_n\}$: a vector of evaluations, one for each sub-game.
$\{e_1, ..., e_n\}$: a set of booleans indicating whether sub-games are fully explored or not, initialized to $\{false, ..., false\}$
$\{d_1, ..., d_n\}$: depth where a revision of a "fully explored" move flag occurred

At each step of the selection (Algorithm 2), the legal moves evaluation is processed in the global game and an expansion is attempted (l.8–9). To try an expansion (Algorithm 3), a random move is played (l.33). Each sub-game is informed of the new sub-state reached. A new transition, and possibly a new node, are added to the sub-tree if necessary (l.36–39). If the transition is already known, i.e. a previously visited action triggers the same transition, the transition is labeled with these different moves which form a *meta-action* [6]. If a legal move triggers an already known transition in each sub-game, it is already evaluated and it is not necessary to test it: this move leads to a combination of already evaluated sub-states. Then, another move is randomly chosen. Playing with decomposed games therefore allows to reduce significantly the search space size. If no legal move leads to an expansion in one of the sub-games, the selection continues.

In a game like *Incredible*, the *Maze* sub-game is quickly fully explored. Then it systematically recommends the sequence of moves leading to the maximum gain allowed by this sub-game. However, ending this sub-game terminates the

---

[3] An informal presentation of MT-MCTS with an example has been published in Journées d'Intelligence Artificielle Fondamentale (JIAF) 2019.

global game prematurely. For a local selection algorithm based on a balance between exploration and exploitation, it takes a large number of visits of the *Maze* terminal move to guide the search towards the exploration of other possible moves (playing in the *Block* sub-game). To alleviate this problem, the terminal legal moves are evaluated (1.10). If a terminal move returns the maximum score possible for the current player, it is always selected (1.12–15), otherwise, the selection continues with the non-terminal moves.

Algorithm 2. `selectionWithExpansion`$(\{s_1, ..., s_n\})$

6: **loop**
7:    $\mathscr{S} := \mathscr{S} \cup \{s_1, ..., s_n\}$
8:    L ← `getLegalMoves`(S)
9:    **if** `expansion`$(\{s_1, ..., s_n\}$, L$)$ **then return** $\{\emptyset, ..., \emptyset\}$
10:   T ← `filterTerminalMoves`(L)
11:   $\{d_1, ..., d_n\}$ ← `checkNotFullyExplored`$(\{s_1, ..., s_n\}$, T$)$
12:   $\{$best, score$\}$ ← `getBestMove`(T)
13:   **if** score = maximum possible evaluation **then**
14:       $\mathscr{M}$ ← $\mathscr{M} \cup$ best
15:       **return** maximum evaluation for each sub-game
16:   explored ← *true*
17:   **for** i in $\{1, ..., n\}$ **do**
18:       **if** $\exists m \in$ L - T: $\neg$ `fullyExplored`(s_i, m) **then**
19:           explored ← *false*
20:       **else**
21:           `fullyExplored`$($last$(\mathscr{S})$, last$(\mathscr{M})) ←$ *true*
22:           e_i ← *true* if $\{s_1, ..., s_n\}$ = *initial_state*
23:   **if** explored but $\exists i : e_i = false$ **then**
24:       **return** mean evaluation for each sub-game
25:   **for** i in $\{1, ..., n\}$ **do**
26:       selected$_i$ ← `localSelectPolicy`$(s_i$, L - T$)$
27:   best ← `globalSelectPolicy`$(\{$selected$_1$, ..., selected$_n\})$
28:   $\mathscr{M}$ ← $\mathscr{M} \cup$ best
29:   S ← `apply`(S, best)
30:   **if** `terminal`(S) **then return** $\{\emptyset, ..., \emptyset\}$
31:   $\{s_1, ..., s_n\}$ ← `getSubstates`(S)

To avoid re-visiting fully explored branches unnecessarily, we flag them to encourage visits to nearby branches. However, in the case of sub-games, a sub-state is not always terminal depending on the rest of the overall state. It is then necessary to develop a specific approach to flag the fully explored branches in sub-trees. We solve this problem by simply revising the labeling during selection and expansion update (1.11,38). If an action was flagged terminal during previous descents in the tree but is not terminal in the current situation, or if a new transition is added, the labeling is revised and revisions are back-propagated along the descent path (1.54–55).

When all legal moves from a sub-state are terminal or fully explored, the previous move is also flagged "fully explored" (1.21). If the current state is the initial state, the whole sub-tree is fully explored (1.22). When the current sub-state is fully explored in each subgame, the mean evaluation is computed and returned (1.24). Otherwise a local selection policy is applied in each sub-game on non-terminal legal moves (1.26). The best move is selected according to a global selection policy (1.27).

Algorithm 3. expansion($\{s_1, ..., s_n\}$, L)

32: **while** L $\neq \emptyset$ **do**
33:     $m \leftarrow$ popRandomMove(L)
34:     S' $\leftarrow$ apply(S, m)
35:     $\{s'_1, ..., s'_n\} \leftarrow$ getSubstates(S')
36:     new_transition $\leftarrow$ *false*
37:     **for** i in $\{1, ..., n\}$ **do**
38:         **if** update(s_i, s'_i, m) **then**   ▷ possible inconsistency is detected here
39:             new_transition $\leftarrow$ *true*
40:     **if** new_transition **then**
41:         $\mathcal{M} \leftarrow \mathcal{M} \cup m$
42:         S $\leftarrow$ S'
43:         **return** *true*
44: **return** *false*

Algorithm 4. simulationWithBackProp($\{val_1, ..., val_n\}$)

45: **if** $\{val_1, ..., val_n\} = \{\emptyset, ..., \emptyset\}$ **then**
46:     **if** ¬terminal(S) **then**
47:         S $\leftarrow$ playoutWithDefaultPolicy(S)
48:     **for** i in $\{1, ..., n\}$ **do**
49:         $val_i \leftarrow$ {global score, max score for sub-game i with 3-valued logic}
50: **for** p in $\{1, ..., \mathcal{M}$.length$\}$ **do**
51:     $\{s_1, ..., s_n\} \leftarrow \mathcal{S}_p$
52:     $m \leftarrow \mathcal{M}_p$
53:     **for** i in $\{1, ..., n\}$ **do**
54:         **if** $p < d_i$ **then**
55:             fullyExplored(s_i, m) $\leftarrow$ *false*
56:         $t \leftarrow$ transition from s_i labeled with m
57:         update(N, n_t, w_t, w_t^{max})   ▷ see local selection policy

---

If the selection ends on a non-evaluated state (Algorithm 4, 1.45), a *playout* is done to reach a terminal state if necessary (1.47), then the state is evaluated (1.49). The details of this evaluation depend on the *local selection policy* and are explained below. The evaluation is back-propagated along the visited path (1.50–57) and the "fully explored" flags are revised if necessary.

The Local Selection Policy chooses the best moves among legal non-terminal moves. The current sub-state is associated with a number of visits N. Each transition t from this sub-state is evaluated by a number of visits n_t and a cumulated score w_t. The local selection returns a set of moves if there exists different transitions with the same evaluation or if the best transition is labeled with several moves.

We investigate different ways to perform this local selection. The first one is a standard application of the UCT policy. However, this approach is not satisfactory because a transition in a sub-game can receive a good evaluation without contributing to the global score: the evaluation was obtained thanks to a move sequence leading to a positive evaluation in another sub-game. Another more troublesome problem occurs in the case of binary scores: the score is always zero until a solution is found. The search is in this case reduced to a breadth first search and is not guided to the right combination of sub-states.

In the GGP framework, a game state is described by a finite set of instantiated fluents, some of which are true. The decomposition partitions these fluents in several groups which represent the sub-games states. In a global terminal state, it is possible to keep only the fluents corresponding to a sub-game, to give an *undefined* value to the other and to evaluate the logic rules of the game with a 3-valued logic. The true or undefined `goal` predicate instances represent the possible scores according to this sub-game state. The maximum `goal` score (*lmax*) corresponds to the maximum potential score that can be obtained if the best possible configuration is found in the other sub-games. The *lmax* score is a maximum indication, the true maximum score may not match exactly because using a 3-valued logic does not guarantee the most accurate information [13]. *lmax* evaluation is nevertheless a valuable indication of the sub-state value. It can be back-propagated in addition to the global score and cumulated in a w_t^{max} variable. For a given transition, w_t/n_t is a global score estimator and w_t^{max}/n_t is a local score estimator. These two estimators can be used in a new policy derived from UCT:

$$U = \alpha \frac{w_t}{n_t} + (1 - \alpha) \frac{w_t^{max}}{n_t} + C * \sqrt{\frac{\log N}{n_t}} \tag{1}$$

with C the constant balancing the exploitation and exploration terms and $\alpha \in [0, 1]$ setting the balance between local and global score estimations.

To avoid going back into already fully explored sub-branches, the transitions corresponding to these branches are excluded from the local selection as long as there are transitions that are not fully explored.

The Global Selection of the best move is made depending on the moves recommended by the sub-games. In serial or synchronous parallel games, the intersection of the recommended move sets is always non-empty. The global selection is then straightforward: a move can be randomly selected in the intersection. However, in a parallel asynchronous game, different sub-games can propose disjoint move sets. It is then necessary to define a policy for the choice of a move at the global level. To define an any game policy, we propose a voting policy. Each sub-game recommending a move brings a vote. In the case of serial or synchronous parallel games, the best moves get as many votes as there are sub-games. In the case of parallel asynchronous games, each move may win a maximum of one vote. Among the moves that received the most votes, those with the highest expected earnings are selected to direct the search towards moves that can lead to the best combination of sub-states. When the game goal is the conjunction of the sub-games goals, this highest expected earning for a move m among T sub-games is the product of the probability of gain in each sub-game s: $\prod_{s=1}^{T} w_m^s / n_m^s$. When the game goal is the disjunction of the sub-games goals, this highest expected earning is the sum of the probability: $\sum_{s=1}^{T} w_m^s / n_m^s$ with w_m^s the cumulated rewards earned during playouts and n_m^s the number of visits of the transition labeled with that move m in sub-game s. If several moves offer the greatest probability of gain, one of them is randomly selected.

4   Experiments

We present here experiments on MT-MCTS[4]. Firstly we evaluate different weighting of our local selection policy and secondly we compare the effectiveness of MT-MCTS against UCT and show that this approach can reduce the overall number of simulations and the solving time. We conducted our experiments on several single-player games: *Incredible*, different grids of *Nonogram* of size 5×5 and 6×6 and *Queens08lg*.

Incredible is an interesting game because it is possible to end the game prematurely with a suboptimal score. It is a usual test bed to evaluate players able to exploit decompositions. *Nonogram* is a logic puzzle in which cells in a grid must be colored or left blank according to numbers placed in front of columns and lines. The score is binary and UCT provides no improvement over depth-first or breadth-first search in this game. *Queens08lg* is on the contrary quickly solved by UCT. It is an *Eight Queens puzzle* in which it is illegal to place a queen in a position where it could capture another queen in one move. The game is over when a queen can no longer be placed. See [Anonymised Author PhD], for more information about these games and a detailed presentation of each Nonogram grid. We decompose these games with the statistical approach proposed by [7]. The decomposition time can vary slightly depending on the simulations done to collect statistical information. For each game and each configuration, we realized 10 tests and present the mean number of playouts and mean time necessary to solve the game. A game is considered solved when a maximum score leaf is found.

The purpose of our first experiment is to compare different values of α in the local selection policy (Eq. 1). We use $C = 0.4$ which allow a good balance between exploration and exploitation in a majority of GGP games. We experimented on

---

[4] The experiments are performed on one core of an Intel Core i7 2,7 GHz with 8Go of 1.6 GHz DDR3.

two games: *Incredible* and *Nonogram "checkerboard"*. The results are presented
in Fig. 1. Using only the global score estimator ($\alpha = 1$) in the local selection
policy does not allow to solve *Nonogram* due to its binary score. The estimated
score is always zero in this game and the randomly chosen moves very unlikely
lead to the right combination of sub-states. On the contrary, the local score
estimator ($\alpha = 0$) allows to guide the search in the sub-trees and solve this
Nonogram in less than 5 s. However, the use of $\alpha = 1$ gives better results on
Incredible while $\alpha = 0$ requires almost twice as much time to solve the game.
By varying α, we notice that a small participation of the local score estimator
($\alpha = 0.75$) allows an even better result in *Incredible*. The weighting $\alpha = 0.25$
seems to allows the fastest resolution of *Nonogram*. However, the importance of
the standard deviations, of the same order of magnitude as the resolution times
or an order below, does not allow to identify a significant effect of the variation
of this weighting. Considering these standard deviations, the resolution times
are similar in the three tests mixing the local and global score estimators. More
experiments will be necessary to verify the influence of unequal weighting of
these two pieces of information.

Nevertheless, the association of both estimators appears desirable to constitute a polyvalent policy. In the following experiments, we used $\alpha = 0.5$ for the
local selection since it gives overall the best result.

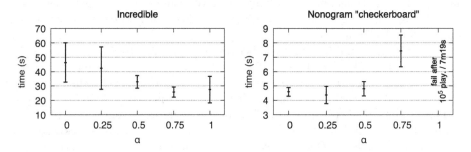

Fig. 1. Mean time on 10 tests to solve *Incredible* and *Nonogram "checkerboard"* with
different values of α in the local selection policy of MT-MCTS (Eq. 1).

In a second experiment (Fig. 2) we compare the effectiveness of MT-MCTS
against UCT in terms of number of playouts and time spent.

On *Incredible*, MT-MCTS is significantly better than UCT in terms of number of playouts necessary to solve the game. MT-MCTS uses twenty times less
playouts but the move selection is significantly longer. In the end, the game is
solved twice as fast.

Comparing our results with [5] (*Fluxplayer*) and [3] (ASP) is difficult because
their approaches are totally different from UCT. *Fluxplayer* takes about 2 h to
solve *Incredible* by computing over 41 million states (compared to 280 thousand

playouts for UCT[5]). Their decomposition method reduces this time to 45 s and 3212 calculated states. Their resolution time is greater than ours although fewer states are computed by their approach. The encoding of *Incredible* in ASP allows game resolution in 6.11 s. This time is reduced to 1.94 s by decomposition, a factor of 3. It should be noted that this approach is optimized for solitary games. Since our approach requires 21 times less playouts for the resolution of the decomposed game, by optimizing the selection step of MT-MCTS, we hope to obtain a similar or even better improvement.

For a simple puzzle like *Queens08lg*, even if the resolution of the decomposed game requires 3 times less playouts, the decomposition time is too important compared to the gain that can be expected in the resolution.

Game	Algo.	# Playouts	Time (decomp.)	σ	# Fail
Incredible	UCT	280158	1m	14.3s	-
	MT-MCTS	13199	32.86s (2.50s)	4.4s	-
Queens08lg	UCT	67	0.01s	<0.1s	-
	MT-MCTS	22	1.30s (1.29s)	<0.1s	-
Nonogram	UCT	432019	31.31s	16.5s	-
"checkerboard"	MT-MCTS	776	4.81s (0.69s)	0.5s	-
Nonogram	UCT	2988921	4m24s	4m8s	-
"G"	MT-MCTS	9344	36.61s (0.77s)	16.45s	-
Nonogram	UCT	4969111	7m20s	3m53s	1 *(after 10^7 play. / 15m17s)*
"tetris"	MT-MCTS	3767	16.36s (1.01s)	6.5s	-
Nonogram	UCT	2552082	3m43s	2m50s	-
"sitdog"	MT-MCTS	5476	26.27s (0.98s)	14.92s	-
Nonogram	UCT	3086172	4m34s	3m9s	-
"iG"	MT-MCTS	10232	28.60s (0.85s)	12.40s	-
Nonogram	UCT	4284872	13m7s	6m41s	4 *(after 10^7 play. / 31m9s)*
"rally"	MT-MCTS	1762	49.26s (2.40s)	3.72s	-
Nonogram	UCT	21438731	1h17m23s	19m16s	8 *(after 5×10^7 play. / 3h4m)*
"cube"	MT-MCTS	358608	1h53m8s (2.75s)	45m7s	-

Fig. 2. Comparison of the resolution times of different games with UCT and MT-MCTS. The columns present the mean number of playouts and the mean time to solve the puzzles (failures excluded) over 10 tests. In parentheses is the time used for decomposition. The σ column indicates the standard deviations. The column "*# Fail*" indicates how many searches were stopped without finding the solution.

On *Nonograms* grids (5×5 and 6×6) MT-MCTS is significantly better than UCT. The resolution is 25 times faster for *"Tetris"*. This gain is not directly related to the number of sub-games identified: in *"Checkerboard"*, which has a larger number of sub-games, the resolution is only 6 times faster. The gain obtained for the resolution speed is directly related to the number of simulations needed: 300 times less for *"iG"* up to 2400 times less for *"rally"* (not to mention the 4 tests where UCT was interrupted after 10^7 playouts without finding any solution). The heavier selection step is largely mitigated here by the significant

---

[5] As each playout results in an expansion of the tree, we can compare the number of playouts with the number of calculated states.

gain in the number of simulations. The time needed to solve *"cube"* with MT-MCTS is 2 h on average. The resolution with UCT is sometimes successful in less than an 2 h, but the majority of tests (8/10) failed to find the solution after 3 h on average.

5   Discussion

During the expansion, some actions are not tested as we already have an evaluation for the resulting sub-states from previous descents in the sub-trees. Therefore some combinations of sub-states may never be visited and splitting the search over several game trees offers in theory no guarantee of convergence toward a solution with an infinite number of playouts. In practice, a good selection policy allows to guide the search to find the right sequence of moves to reach the right combination of sub-states.

The problem in the GGP domain is that the optimal policy depends on the structure of the game. For example, excluding fully explored branches in local selection can quickly guide to the solution in *Incredible* as it avoids re-exploring a path leading to a suboptimal score. However, it can delay the resolution in a game like *Nonogram* where playing the already known good moves would have allowed to color some cells and guide the discovery of the following good moves.

Despite this delay, MT-MCTS is still more efficient than UCT on *Nonogram*. However, as an interesting specific combination of sub-states can remain unexplored for a long time, we assume that a fixed value for α in our policy may not be as effective on all GGP games. Further research is therefore needed. Many different policies could be considered to allow to go down rarely in the fully explored branches and to improve the selection step of MT-MCTS. Finding a policy for MT-MCTS that is proven effective on all games is an interesting open problem.

Nonogram naturally presents a composition of rules in rows and columns. The structure of MT-MCTS allows to explore a game decomposed in this way. Another avenue of research to consider is then the exploitation of different overlapping sub-games and, more generally, of non-disjoint sub-games.

Our version of MT-MCTS does not consider all transpositions to avoid games with cycles. Fifty-five percent of GGP games use a *stepper*, the development of a specific selection policy to take advantage of transpositions in games containing cycles is therefore an open and interesting search track that could significantly improve the level of GGP players.

6   Conclusion

In this paper we proposed an extension of MCTS to search in several trees representing the different parts of a decomposed problem. We tested this idea on several single player games in the *General Game Playing* domain. Playing with decomposed games allows to hope for a real change of scale in their resolution speed. Our tests with a weighted selection policy give promising results: the games are solved from 2 times (*Incredible*) to 25 times faster (*Nonogram*). Multiple Tree MCTS (MT-MCTS) can be extended to multi-player games such as conventional MCTS approaches and also allows non-independent sub-games to be exploited.

The new MT-MCTS approach opens different research tracks: the development of a selection policy efficient for the different types of compound games, the support of the specific case of games with cycles using a *stepper*, playing with overlapping sub-games and even the exploitation of incomplete or imperfect decompositions.

References

1. Browne, C.B., et al.: A survey of Monte Carlo Tree search methods. IEEE Trans. Comput. Intell. AI Games **4**(1), 1–43 (2012)
2. Cazenave, T.: Monte-Carlo approximation of temperature. Games Chance **4**(63), 41–45 (2015)
3. Cerexhe, T., Rajaratnam, D., Saffidine, A., Thielscher, M.: A systematic solution to the (de-)composition problem in general game playing. In: Proceedings of the European Conference on Artificial Intelligence (ECAI) (2014)
4. Cox, E., Schkufza, E., Madsen, R., Genesereth, M.: Factoring general games using propositional automata. In: Proceedings of the IJCAI-09 Workshop on General Game Playing (GIGA 2009), pp. 13–20 (2009)
5. Günther, M., Schiffel, S., Thielscher, M.: Factoring general games. In: Proceedings of the IJCAI-09 Workshop on General Game Playing (GIGA 2009), pp. 27–33 (2009)
6. Hufschmitt, A., Méhat, J., Vittaut, J.N.: A general approach of game description decomposition for general game playing. In: Proceedings of the IJCAI-16 Workshop on General Game Playing (GIGA 2016), pp. 23–29 (2016)
7. Hufschmitt, A., Vittaut, J.N., Jouandeau, N.: Statistical GGP games decomposition. In: Proceedings of the IJCAI-18 Workshop on Computer Games (CGW 2018), pp. 1–19 (2018)
8. Kishimoto, A., Müller, M.: A general solution to the graph history interaction problem. In: Nineteenth National Conference on Artificial Intelligence (AAAI 2004), San Jose, CA, pp. 644–649 (2004)
9. Kocsis, L., Szepesvári, C.: Bandit based Monte-Carlo planning. In: Fürnkranz, J., Scheffer, T., Spiliopoulou, M. (eds.) ECML 2006. LNCS (LNAI), vol. 4212, pp. 282–293. Springer, Heidelberg (2006). https://doi.org/10.1007/11871842_29
10. Love, N., Hinrichs, T., Haley, D., Schkufza, E., Genesereth, M.: General game playing: game description language specification. Tech. rep. LG-2006-01, Stanford University, January 2006
11. Mehat, J., Cazenave, T.: Combining UCT and nested monte carlo search for single-player general game playing. IEEE Trans. Comput. Intell. AI Games **2**(4), 271–277 (2011)
12. Palay, A.: Searching With Probabilities. Research Notes in Artificial Intelligence Series, Pitman Advanced Publishing Program (1985)
13. Reps, T.W., Loginov, A., Sagiv, S.: Semantic minimization of 3-valued propositional formulae. In: 17th IEEE Symposium on Logic in Computer Science (LICS 2002), 22–25 July 2002, Copenhagen, Denmark, Proceedings, p. 40 (2002)

14. Saffidine, A., Méhat, J., Cazenave, T.: UCD: upper confidence bound for rooted directed acyclic graphs. In: TAAI 2010, Piscataway, NJ, pp. 467–473. IEEE (2010)
15. Silver, D., et al.: Mastering the game of go without human knowledge. Nature **550**, 354–359 (2017)
16. Winands, M.H.M., Björnsson, Y., Saito, J.-T.: Monte-Carlo tree search solver. In: van den Herik, H.J., Xu, X., Ma, Z., Winands, M.H.M. (eds.) CG 2008. LNCS, vol. 5131, pp. 25–36. Springer, Heidelberg (2008). https://doi.org/10.1007/978-3-540-87608-3_3
17. Zhao, D., Schiffel, S., Thielscher, M.: Decomposition of multi-player games. In: Nicholson, A., Li, X. (eds.) AI 2009. LNCS (LNAI), vol. 5866, pp. 475–484. Springer, Heidelberg (2009). https://doi.org/10.1007/978-3-642-10439-8_48

On Efficiency of Fully Probing Mechanisms in Nonogram Solving Algorithm

Yan-Rong Guo[1], Wei-Chiao Huang[1], Jia-Jun Yeh[1], Hsi-Ya Chang[2], Lung-Pin Chen[3], and Kuo-Chan Huang[1(✉)]

[1] National Taichung University of Education, No. 140, Minsheng Road,
West District, Taichung 40306, Taiwan
`acs103149@gmail.com`, `erica.ttc@gmail.com`, `s000032001@gmail.com`,
`kchuang@mail.ntcu.edu.tw`
[2] National Center for High-Performance Computing, Hsinchu, Taiwan
`9203117@narlabs.org.tw`
[3] Tunghai University, Taichung, Taiwan
`lbchen@thu.edu.tw`

Abstract. Fully probing plays an important role in the nonogram solving algorithm developed by Wu *et al.*, whose implementation, named LalaFrogKK, has won several nonogram tournaments since 2011. Different fully probing methods affect the overall nonogram solving performance greatly as shown in previous studies. In this paper, we explore fully probing efficiency from different aspects and evaluate its impact on the performance of solving an entire nonogram puzzle. In the exploration, we found several critical factors influencing fully probing efficiency greatly, i.e. re-probing policy, probing sequence, and computational overhead. Taking these critical factors in account, we developed several new fully probing mechanisms to improve nonogram solving performance. Experimental results based on the puzzles of previous nonogram tournaments show that our new fully probing methods have the potential to improve the speed of solving nonogram puzzles significantly.

Keywords: Nonogram · Line solving · Propagation · Fully probing · Backtracking

1 Introduction

A nonogram puzzle [14] looks like Fig. 1, which is a kind of picture logic puzzles with clues given as row and column constraints. Solving a nonogram puzzle is to paint each pixel in black or white under the given row and column constraints. Figure 1 illustrates a solved nonogram puzzle, represented by a 5 × 5 grid with its row clues given next to the leftmost column and column clues showed above the top row. Each clue is an integer indicating the required length of a segment of consecutive black pixels which should appear in the corresponding row or column in the solution. Moreover, segments of black pixels implied by consecutive clues should be separated by at least one white

T. Cazenave et al. (Eds.): ACG 2019, LNCS 12516, pp. 119–130, 2020.
https://doi.org/10.1007/978-3-030-65883-0_10

pixel. For example, in Fig. 1 the clues of the third row, 2 1, indicate that the row in the solution should have exactly two segments of consecutive black pixels, starting from the left, where the length of the first segment is 2 and that of the second segment is 1. It is possible that a nonogram puzzle has more than one solution all satisfying the constraints.

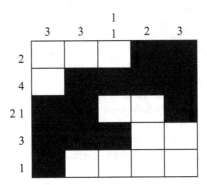

Fig. 1. A solved 5 × 5 nonogram puzzle

Solving a nonogram puzzle has been shown to be an NP-complete problem [20], and received a lot of research attention [1, 8, 24]. Wu *et al.* [23] proposed an efficient approach to solving nonograms, featuring a fast dynamic programming method for line solving, followed by fully probing (FP) methods to solve as many pixels as possible before the final backtracking stage. A nonogram solver developed based on the approach in [23], named LalaFrogKK [13], has won several nonogram tournaments [12].

In general, a nonogram puzzle might have more than one solution. Regarding each pixel in a nonogram puzzle, it might either have only one choice for its value or could have two alternative choices, depending on the clues of the puzzle and those pixels already being set values. For the pixels potentially having alternative choices, their values might have interrelationship and depend on each other under the row and column constraints. The algorithm in [23] adopts a conservative approach to solving a nonogram puzzle. It tries to paint those pixels with only once choice first in the earlier stages, and then solve the other pixels with two alternative choices by figuring out their interrelationship.

Line solving is the basic unit of the nonogram solving approach in [23], which treats a line, i.e. a single row or a column, at a time, trying to determine the values of those pixels with only one choice in the line based on the information of line clues and the values of the already solved pixels in the same line. Line solving in [23] is based on the recurrence definitions of Fix and Paint functions. A dynamic programming method was proposed in [23] to improve the speed of computing the recursive functions.

Based on line solving, an iterative procedure, Propagate, was proposed to paint as many one-choice pixels as possible for the entire grid by checking potential conflicts between specific values of pixels and clues. Propagate in [23] is conservative in the sense that it paints a pixel into a specific value, i.e. 0 for white pixels and 1 for black pixels, only when it could confirm that the other value would result in a conflict with some line clues. Therefore, during the nonogram solving process, a temporary unknown value for a pixel doesn't necessarily mean that it could have two alternative choices in the puzzle.

It is only a sign that we have not found any evidence for it to be a single-choice pixel yet. Later in the solving process, it could finally turn out to be either a single-choice pixel or a two-choice pixel.

When no more pixels can be painted after the Propagate procedure, a common approach in previous works is to employ backtracking to search all possible solutions. On the other hand, in [23] fully probing was proposed to be used before backtracking. It attempts to guess a value, i.e. 0 or 1, on each pixel to see whether or not more pixels can then be painted by a following re-run of the Propagate procedure. Fully probing is different from backtracking in that it is not recursive. In contrast to Propagate, fully probing is more aggressive since it guesses a value first and then checks whether or not that guess would result in a conflict with clues on other lines, while Propagate would determine a pixel's value only when it could immediately find the conflict between a specific value and the clues right on the line containing the pixel.

Since fully probing is also used in later backtracking stage, it plays an important role in the nonogram solving approach in [23]. In our previous work [9], we proposed a new fully probing method, called selective_repeat, to paint new pixels with less efforts at the algorithmic level. Experimental results in [9] showed that selective_repeat can paint almost as many pixels as the FP1 method in [23] but with relatively much less probing activities, resulting in faster nonogram solving speed. However, for very few puzzles in the experiments, selective_repeat would paint one less pixel than FP1. This finding motivated our research work presented in this paper. We investigated the efficiency issues of fully probing from different aspects at both algorithmic and implementation levels. Three factors were found to have critical influence on fully probing efficiency, i.e. re-probing policy, probing sequence, and computational overhead. Based on the research results, we present several new fully probing mechanisms to improve nonogram solving performance further in this paper. Experimental results show that our new fully probing methods could improve the speed of solving nonogram puzzles significantly, ranging from 11% to 99%.

2 Related Work

Many researchers have tried to tackle the nonogram solving problem from different angles. Translating nonogram solving into other classical problems is one of the major approaches. For example, Bosch translated it into an integer linear programming problem for solution in [4]. On the other hand, Faase [8] translated it into an exact cover problem and then used Knuth's dancing-links method [11] to solve it. Unfortunately, nonogram puzzles usually cannot be solved efficiently by these kinds of problem transformations, especially for large nonogram puzzles.

Meta-heuristic algorithms have been used by some researchers to solve nonogram puzzles. For example, Wiggers [21] used a genetic algorithm to solve nonogram puzzles, while Batenburg [1] proposed an evolutionary algorithm for nonogram solving. Approaches of this kind can also be found in [17–19]. However, one potential drawback of these approaches is that they cannot guarantee to solve a puzzle.

Another kind of nonogram solving approaches are based on logical rules. For example, Yu et al. [24] developed an algorithm which tries to solve nonogram puzzles based on specific logical rules in the beginning, and then use backtracking to find the solutions of unpainted pixels aided with the same set of logical rules for improving search efficiency. Jing [10] divided nonogram solving into two parts. In the first part, logical rules are used to paint some pixels first. The remaining unpainted pixels are then solved using a depth first search method enhanced by the branch and bound mechanism in the second part. These logical rules are usually focused on painting as many pixels as possible in a line-by-line manner. There are also other line painting methods proposed in [15, 16], which are simpler, however, cannot guarantee painting as many pixels as in [10, 24] for each row or column.

For line solving to paint as many pixels as possible for each row or column, Batenburg and Kosters [3] adopted a dynamic programming approach instead of logical rules. In general, their approach paints more pixels than those in [15, 16, 24], and thus result in faster nonogram solving speed. Moreover, they used a 2-Satisfiability method to paint more pixels before the backtracking stage. [22] presents information of many nonogram solvers. Generating nonogram puzzles is also an important issue for conducting nonogram solving research. Several nonogram-generating algorithms are discussed in [2].

The nonogram solving approach in [23] consists of three major parts: propagation based on line solving, fully probing, and backtracking. They developed a faster dynamic programming method for line solving based on previous research in [3]. Three fully probing methods were also proposed for painting more pixels before the backtracking part. In addition to painting more pixels after the propagation part, fully probing also provides the later backtracking stage with useful information about which pixel to guess first in order to find the solution of a nonogram puzzle more quickly. Many nonogram solvers participating recent tournaments are based on the approach in [23] and try to make some enhancements, such as [5–7, 9].

3   Fully Probing Efficiency from Different Aspects

The nonogram solving approach in [23] has three major components: Propagate() for conservative pixel painting, FP() for painting pixels in a more aggressive way, and Backtracking() for brute-force depth-first search with some smart guessing techniques. However, the three components are not executed one by one. They exhibit a particular computational structure where Backtracking() invokes FP() and FP() would call Propagate().Therefore, FP() plays an major role in the entire nonogram solving process and affects the overall performance greatly. That's why we choose fully probing as the research topic for improving nonogram solving efficiency further. The entire nonogram solving process is as follows. It tries to paint as many pixels as possible first by conducting the fully probing method at line 3, which would invoke Propagate(). After that, all unsolved pixels are painted by the recursive backtracking procedure.

Nonogram solving algorithm in [23]

1. Initialize G to be the grid representing the solution of a nonogram puzzle with all pixels being unsolved. G is also accompanied by the row and column clues of the nonogram puzzle.

2. Initialize all $G_{p,0}$ and $G_{p,1}$ for each pixel p, where p is guessed to be 0 and 1, respectively.
3. FP(G);//conduct the fully probing method in [23]
4. **if** (status(G) is CONFLICT or SOLVED) **then return**;
5. Backtracking(G);

The idea of fully probing in [23] is to make guesses for all unpainted pixels. Each guess would result in a new partial solution grid. Then, it performs Propagate on each resultant partial solution grid. The following shows the first proposed fully probing method in [23], named FP1.

Procedure FP(G) //the FP1 fully probing method in [23]
    1.   **repeat**
    2.   PROPAGATE(G)
    3.   **if** (status(G) is CONFLICT or SOLVED) **then return**
    4.   **for** (each unpainted pixel p in G) **do**
    5.   Update $G_{p,0}$ and $G_{p,1}$ with newly painted pixels in G
    6.   PROBE(p) //invoking Propagate() on $G_{p,0}$ and $G_{p,1}$ inside to paint more pixels
    7.   **if** (status(G) is CONFLICT or SOLVED) **then return**
    8.   **if** (status(G) is PAINTED) **then break**
    9.   **end for**
   10.   **until** (no more pixels can be painted)
end procedure

In the following, we investigate into three potential factors that affect fully probing efficiency, respectively.

3.1 Re-probing Policy

In the fulling probing method proposed in [23], as shown in the definition of FP() in the previous section, line 8 is a key part which enforces re-probing all the unsolved but already probed pixels by the break statements once the most recent probing successfully paints new pixels. This mechanism is important to make sure one run of FP() could paint as many pixels as possible. However, we found that the goal is achieved at the cost of inefficient fully probing because many re-probing activities are not necessary and result in no more pixels painted.

We explored this issue and proposed several new re-probing polices, which all outperformed the original fully probing method in [23] significantly, in our previous work [9]. Selective_repeat was the best re-probing policy in [9]. However, our goal, i.e. painting as many pixels as the original fully probing in [23] with less probing activities, was not achieved completely by the selective_repeat policy since for very few puzzles its fully probing result is one less pixel than that of the original method in [23]. The unachieved goal motivated our research work in this paper.

In the following, we evaluate three existing re-probing policies, original FP in [23], one_run [9], and selective_repeat [9], together with a new enhancement mechanism to selective_repeat, and investigate the issue of fully probing efficiency at an algorithmic level, measuring how many pixels can be painted with how many times of pixel probing.

The following describes the re-probing policies to be evaluated in this section except the original fully probing method in [23].

- **One_run**. In this policy, no pixel re-probing would be conducted. Each unsolved pixel would be probed for only once in each FP() invocation.
- **Selective_repeat**. In this policy, fully probing is conducted iteratively. At the first iteration, all unsolved pixels would be probed. In later iterations, it tries to re-probe only those pixels promising to paint new pixels. An unsolved pixel is regarded as promising for re-probing only when it is located at the row or column crossing on a pixel newly painted at last iteration. The iterative fully probing process would continue until no new pixels are painted after an entire iteration.
- **Selective_repeat_with_final_check**. This enhancement adds an iteration of probing all unsolved pixels for once when the selective_repeat fully probing procedure paints no more pixels after an iteration. The fully probing procedure would come to an end only when this added iteration still cannot paint any new pixels. Otherwise, the procedure would continue following the selective_repeat policy.

Table 1 compares the four re-probing policies with three puzzles selected from the 1000 TAAI 2014 nonogram tournament puzzles in terms of total pixels painted, how many times of probing performed, how many times of probing in vain (i.e. painting no more pixels), and how many pixels can be painted by a probing in average. The data were measured after the first run of fully probing when solving a nonogram puzzle. The experimental results show that both FP in [23] and selective_repeat_with_final_check could paint most pixels for all the three puzzles. In terms of efficiency, the new selective_repeat_with_final_check policy is obviously better than FP in [23] since it painted the same number of pixels with much less times of probing. Selective_repeat also painted the most number of pixels as those two policies for two puzzles, but painted one less pixel for the other puzzle. However, it required much less times of probing to paint those pixels, resulting in a much higher value of painted pixels per probing than FP in [23] and selective_repeat_with_final_check. One_run painted the least number of pixels in each puzzle, but also conducted the fewest times of probing. It achieves the highest value of painted pixels per probing, i.e. most efficient at the algorithmic level among all re-probing policies.

3.2 Probing Sequence

The fully probing method in [23] doesn't describe any specific sequence for probing unsolved pixels. One natural and common approach is to probe the unsolved pixels row by row starting from the left top corner in the puzzle grid, which was used in LalaFrogKK and our previous solvers. In our previous work [9], we pointed out that probing sequence could also be an algorithm-level issue with experimental evidence. However, we didn't offer any specific better probing sequence in [9]. In the following, we propose a new clue-based probing mechanism and compare it to the commonly used one.

The clue-based probing mechanism assigns a priority value to each unsolved pixel, and then probes all unsolved pixels in the specific order determined by their priority values. The basic idea of the clue-based probing mechanism is that the importance of

Table 1. Comparison of re-probing policies

	Painted pixels	Times of probing	Times of probing in vain	Painted pixels per probing
Puzzle #50				
FP in (Wu et al. 2013)	569	299	271	1.903
One_run (Huang et al. 2018)	567	91	65	6.231
Selective_repeat (Huang et al. 2018)	569	161	134	3.534
Selective_repeat_with_final_check	569	204	177	2.789
Puzzle #100				
FP in (Wu et al. 2013)	607	5365	5295	0.113
One_run (Huang et al. 2018)	375	292	266	1.284
Selective_repeat (Huang et al. 2018)	607	433	387	1.402
Selective_repeat_with_final_check	607	433	387	1.402
Puzzle #309				
FP in (Wu et al. 2013)	191	7398	7352	0.026
One_run (Huang et al. 2018)	160	493	464	0.325
Selective_repeat (Huang et al. 2018)	190	1520	1475	0.125
Selective_repeat_with_final_check	191	2278	2232	0.084

an unsolved pixel, i.e. how promising it can lead to more pixels being painted, is related to the clues of the row and column crossing on it. In the following, the priority value of each unsolved pixel is calculated by summing up the clues of the row and column crossing on it, and the probing sequence is in the decreasing order of priority values. The assumption is that unsolved pixels on the rows and columns with constraints of more black pixels are more promising to paint more new pixels in the next probing. Table 2 evaluates the new clue-based probing mechanism by comparing it to the One_run [9] fully probing method. Both methods probe each unsolved pixel for at most once in a single run of fully probing, but One_run adopts a simple row-wise probing sequence starting from the left-top corner of the puzzle grid.

The experimental results in Table 2 confirm that probing sequence is an algorithm-level issue affecting performance significantly, and show that clue-based probing is promising to paint more pixels than the original method in a single run of fully probing. In addition to painting more pixels, clue-based probing also results in less times of probing in vain and thus more painted pixels per probing in average, improving fully probing efficiency. The higher efficiency of the clue-based probing mechanism also leads to its faster speed than One_run for solving the three puzzles.

Table 2. Comparison of probing sequences

	Painted pixels	Times of probing	Times of probing in vain	Painted pixels per probing	Execution time
Puzzle #50					
One_run (Huang et al. 2018)	567	91	65	6.231	0.103
Clue_based probing	569	83	62	**6.855**	**0.041**
Puzzle #100					
One_run (Huang et al. 2018)	375	292	266	1.284	0.043
Clue_based probing	607	246	197	**2.467**	**0.037**
Puzzle #309					
One_run (Huang et al. 2018)	160	493	464	0.325	0.209
Clue_based probing	170	496	452	**0.343**	**0.204**

3.3 Computational Overhead

The methods discussed in the previous sections can be divided into two categories. The first category, e.g. the clue-based probing mechanism, probes each unsolved pixel for just once, while the second category, e.g. selective_repeat [9], would conduct re-probing on all or some unsolved pixels. In general, fully probing methods of the second category could paint more pixels as shown in Table 1, however, at the cost of larger computational overheads because of more times of probing in vain. The computational overheads of fully probing methods should be taken seriously because fully probing would also be repeatedly conducted in the later backtracking stage, not just run for once.

Table 3 evaluates the effects of such computational overheads on the speed of solving a nonogram puzzle. For easier puzzles, e.g. puzzles #50 and #100 in Table 3, more pixels could be painted after the first run of fully probing and thus the computational overheads incurred by the methods of the second category don't hurt the overall performance of nonogram solving since only fewer unsolved pixels are left for the backtracking stage. Therefore, selective_repeat outperforms the clue-based probing method in both puzzles #50 and 100 in terms of execution time. On the other hand, for more difficult puzzles, such as puzzles #799 and #990 in Table 3, only very few pixels can be painted after the first run of fully probing. Since many unsolved pixels are left to the backtracking stage, requiring many times of fully probing at that stage, the computational overheads of fully probing methods could have a crucial influence on the overall nonogram solving performance. Therefore, as shown in Table 3, the clue-based probing mechanism with less computational overheads outperforms selective_repeat in both puzzles #799 and #990. The results demonstrate that it is difficult to have a single fully probing method which can achieve the best performance for all nonogram puzzles.

Table 3. Effects of computational overhead

	Painted pixels	Times of probing	Execution time
Puzzle #50			
Clue_based probing	569	83	0.041
Selective_repeat (Huang et al. 2018)	569	161	**0.039**
Puzzle #100			
Clue_based probing	607	246	0.037
Selective_repeat (Huang et al. 2018)	607	433	**0.024**
Puzzle #799			
Clue_based probing	1	624	**0.102**
Selective_repeat (Huang et al. 2018)	1	672	0.158
Puzzle #990			
Clue_based probing	3	623	**0.116**
Selective_repeat (Huang et al. 2018)	3	717	0.180

4 New Fully Probing Methods and Performance Evaluation

Based on the investigation and experimental results of three crucial factors affecting fully probing efficiency in Sect. 3, we present and evaluate three new fully probing methods in the following.

- **Clue-based fully probing**. This is exactly the clue-based method evaluated in Sect. 3.2, which would probe each unsolved pixel for just once and adopt a new priority-based probing sequence to improve efficiency.
- **Selective_repeat_and_clue_based probing**. This method tries to take into consideration the effects of both re-probing policies and probing sequence. Therefore, it integrates the selective_repeat re-probing policy and the clue-based probing sequence. In this method, each unsolved pixel is possible to be probed for more than one time. The method conducts fully probing in an iterative manner. At each iteration, only some unsolved pixels would be selected for re-probing based on the selective_repeat policy, and the clue-based probing sequence is used to perform probing.
- **Adaptive_FP**. This method aims to make an appropriate balance between effectiveness, i.e. painted pixels, and computational overheads. In the former part of the

nonorgam solving process, it adopts a method which incurs a higher computational overhead, but is promising to paint more pixels in each run of fully probing. On the other hand, in the later part of the nonorgam solving process, it focuses on overhead reduction, and thus uses a method painting each unsolved pixel for just once. The separation between the former and later part of a nonogram solving process is controlled by a parameter defined in terms of the times of fully probing already performed. In the following experiments, Adaptive_FP adopts selective_repeat_and_clue_based probing for the former part and One_run for the later part.

The experiments were conducted on a laptop computer running 64-bit Microsoft Windows and containing an Intel i5-8250U CPU and 12 GB memory. The nonogram solvers were compiled in Dev-C++ 5.11 which is equipped with TDM-GCC 4.9.2. Table 4 evaluates the above three fully probing methods with three different sets of nonogram tournament puzzles, i.e. TAAI 2014, TCGA 2017, TAAI 2018, and compares them to three previous methods in [9, 23]. There are 1000 25 × 25 puzzles for each tournament. Table 4 presents the average execution time of a single puzzle in seconds and the standard deviation among 1000 puzzles for each method. The experimental results show that all the three new fully probing methods have the potential to outperform the three existing methods in [9, 23]. Moreover, for each set of puzzles either the proposed clue-based probing or adaptive FP method achieves the best performance.

Table 4. Performance evaluation of fully probing methods in seconds

	TAAI 2014		TCGA 2017		TAAI 2018	
	Average execution time	Standard deviation	Average execution time	Standard deviation	Average execution time	Standard deviation
FP in (Wu et al., 2013)	1.700674	10.60825	2.62176	23.94344	2.359404	24.42377
One_run (Huang et. al., 2018)	1.260429	21.39171	0.991373	9.881388	1.44393	16.50721
Selective_repeat (Huang et. al., 2018)	0.721714	4.409144	0.986693	9.764836	1.378681	16.76851
Clue_based probing	0.937931	6.103683	**0.89477**	**5.813115**	**0.430268**	**2.775844**
Selective_repeat_and_clue_based probing	0.676481	4.267752	1.02144	10.73068	0.937244	9.727025
Adaptive_FP	**0.64753**	**4.059608**	0.947791	9.487506	0.899915	9.5812

The above experimental results also indicate an interesting finding that the performance of a fully probing method could vary for different set of puzzles, and thus no one method can always achieve the best performance across all puzzles. As an example, Table 5 shows the execution time of two puzzles in TAAI 2014 tournament required by different fully probing methods. One_run [9] outperforms the other methods in solving puzzle #670, but results in the worst performance among all methods when solving puzzle #961. For both puzzles, the performance difference between the best and the worst method is extraordinary large, demonstrating that each method has potential drawback

Table 5. Execution time (sec.) of two puzzles by different fully probing methods

	#670	#961
FP in (Wu et al. 2013)	150.871	0.110
One_run (Huang et al. 2018)	**6.077**	665.990
Selective_repeat (Huang et al. 2018)	60.376	**0.031**
Selective_repeat_and_clue_based probing	58.517	0.047
Adaptive_FP	60.889	**0.031**

for specific kinds of puzzles. We believe that is because different methods lead to different partial solutions after the first run of fully probing, and different characteristics in the partial solutions would affect the required times of the following backtracking greatly. Investigation into this phenomenon is a promising future research direction for developing new methods to further improve the overall performance of nonogram solving.

5   Conclusions and Future Work

This paper presents our research on exploring the issues of fully probing efficiency in solving nonogram puzzles based on the framework proposed in [23]. We found that three critical factors influence fully probing efficiency greatly, i.e., re-probing policy, probing sequence, and computational overhead. Experimental evaluations of the three factors are presented and new probing mechanisms are proposed considering specific factors. Based on such results, three new fully probing methods are presented and evaluated with nonogram puzzles of three different tournaments. Compared to previous methods in [9, 23], the new fully probing methods have the potential to achieve better performance, resulting in performance improvement ranging from 11% to 99% for puzzles of different tournaments.

The experimental results also indicate a promising future research direction, investigating and resolving the inconsistent performance of a fully probing method across different nonogram puzzles. Backtracking process would be the potential target worth further investigation and improvement. A good starting point is to figure out the influence of the characteristics of the partial solution after each fully probing on the following backtracking performance. New search and backtracking methods taking into consideration the characteristics of partial solutions could be developed to resolve the inconsistent performance issue and improve the overall nonogram solving performance further.

References

1. Batenburg, K.J.: A network flow algorithm for reconstructing binary images from dis- crete X-rays. J. Math. Imag. Vis. **27**(2), 175–191 (2007)
2. Batenburg, K.J., Henstra, S., Kosters, W.A., Palenstijn, W.J.: Constructing simple nonograms of varying difficulty. Pure Math. Appl. **20**, 1–15 (2009)

3. Batenburg, K.J., Kosters, W.A.: Solving nonograms by combining relaxations. Pattern Recogn. **42**(8), 1672–1683 (2009)
4. Bosch, R.A.: Painting by numbers. Optima **65**, 16–17 (2001)
5. Chen, L.P., Huang, K.C.: Solving nonogram puzzles by using group-based fully probing. Comput. Games Assoc. J. **40**(4), 387–396 (2018)
6. Chen, L.P., Hung, C.Y., Liu, Y.C.: A new simplified line solver for nonogram puzzle games. In: TCGA Computer Game Workshop (TCGA2017) (2017)
7. Chen, Y.C., Lin, S.S.: A fast nonogram solver that won the TAAI 2017 and ICGA 2018 tournaments. Comput. Games Assoc. J. 1–13 (2019, pre-press)
8. Faase, F.: Nonogram to exact cover, 31 March 2018 (2009). http://www.iwriteiam.nl/D0906.html#28
9. Huang, K.C., Yeh, J.J., Huang, W.C., Guo, Y.R.: Exploring effects of fully probing sequence on solving nonogram puzzles. Comput. Games Assoc. J. **40**(4), 397–405 (2018)
10. Jing, M.Q.: Solving Japanese puzzles with logical rules and depth first search algorithm. In: International Conference on Machine Learning and Cybernetics, pp. 2962–2967 (2009)
11. Knuth, D.E.: Dancing links. In: Millennial Perspectives in Computer Science: the Oxford-Microsoft Symposium in Honour of Sir Tony Hoare (Cornerstones of Computing), Basingstoke, U.K., pp. 187–214. Palgrave (1999)
12. Lin, H.H., Sun, D.J., Wu, I.C., Yen, S.J.: The 2011 TAAI computer-game tournaments. Comput. Games Assoc. J. **34**(1), 51–54 (2011)
13. LalaFrogKK, 11 May 2019 (2015). http://java.csie.nctu.edu.tw/~icwu/aigames/LalaFrogKK.html
14. Nonogram: Wikipedia, the free encyclopedia, 11 May 2019 (2017). http://en.wikipedia.org/wiki/Nonogram
15. Olšák, M., Olšák, P.: Griddlers solver, 11 May 2019 (2003). http://www.olsak.net/grid.html#English
16. Simpson, S.: Nonogram solver, 11 May 2019. http://www.lancaster.ac.uk/~simpsons/software/pkg-nonowimp.htmlz.en-GB
17. Tsai, J.: Solving Japanese nonograms by taguchi-based genetic algorithm. Appl. Intell. **37**(3), 405–419 (2012)
18. Tsai, J., Chou, P.: Solving Japanese puzzles by genetic algorithms. In: International Conference on Machine Learning and Cybernetics, pp. 785–788 (2011)
19. Tsai, J., Chou, P., Fang, J.: Learning intelligent genetic algorithms using Japanese nonograms. IEEE Trans. Educ. **55**(2), 164–168 (2012)
20. Ueda, N., Nagao, T.: NP-completeness results for nonogram via parsimonious reductions. Tech. rep. TR96-0008, Department Computer Science, Tokyo Institute Technology, Tokyo, Japan (1996)
21. Wiggers, W.A.: A comparison of a genetic algorithm and a depth first search algorithm applied to Japanese nonograms. In: Twenty Student Conference IT, pp. 1–6 (2004)
22. Wolter, J.: The 'Pbnsolve' paint-by-number puzzle solver, 11 May 2019 (2012). http://webpbn.com/pbnsolve.html
23. Wu, I.C., et al.: An efficient approach to solving nonograms. IEEE Trans. Comput. Intell. AI Game **5**(3), 251–264 (2013)
24. Yu, C.H., Lee, H.L., Chen, L.H.: An efficient algorithm for solving nonograms. Appl. Intell. **35**(1), 18–31 (2009)

Net2Net Extension for the AlphaGo Zero Algorithm

Hsiao-Chung Hsieh, Ti-Rong Wu, Ting-Han Wei, and I-Chen Wu[⊠]

Department of Computer Science, National Chiao Tung University, 1001 University
Road, Hsinchu, Taiwan, ROC
{michael81420,kds285,ting,icwu}@aigames.nctu.edu.tw

Abstract. The number of residual network blocks in a computer Go
program following the AlphaGo Zero algorithm is one of the key factors
to the program's playing strength. In this paper, we propose a method
to deepen the residual network without reducing performance. Next, as
self-play tends to be the most time-consuming part of AlphaGo Zero
training, we demonstrate how it is possible to continue training on this
deepened residual network using the self-play records generated by the
original network (for time saving). The deepening process is performed
by inserting new layers into the original network. We present in this
paper three insertion schemes based on the concept behind Net2Net.
Lastly, of the many different ways to sample the previously generated
self-play records, we propose two methods so that the deepened network
can continue the training process. In our experiment on the extension
from 20 residual blocks to 40 residual blocks for 9×9 Go, the results
show that the best performing extension scheme is able to obtain 61.69%
win rate against the unextended player (20 blocks) while greatly saving
the time for self-play.

Keywords: AlphaGo Zero · Deep learning · Net2Net

1 Introduction

Since AlphaGo Zero's [7] recent achievement of reaching superhuman level in
Go, there have been numerous projects to reproduce or analyze its core algorithm, such as Facebook AI Research's ELF OpenGo [9], the crowd-sourced
Leela Zero [6], and CGI [10]. The AlphaGo Zero algorithm works by training
deep convolution neural networks (CNNs) using self-play game records, which

H.-C. Hsieh and T.-R. Wu–Equal contribution.
This research is partially supported by the Ministry of Science and Technology
(MOST) under Grant Number MOST 107-2634-F-009-011 and MOST 108-2634-F-009-
011 through Pervasive Artificial Intelligence Research (PAIR) Labs, Taiwan and also
partially supported by the Industrial Technology Research Institute (ITRI) of Taiwan
under Grant Number B5-10804-HQ-01. The computing resource is partially supported
by national center for high-performance computing (NCHC).

© Springer Nature Switzerland AG 2020
T. Cazenave et al. (Eds.): ACG 2019, LNCS 12516, pp. 131–142, 2020.
https://doi.org/10.1007/978-3-030-65883-0_11

requires a large amount of computing resources. During the prototyping process, a common approach is to train a relatively small network, say, 20 residual blocks [2], to ensure that the chosen hyper-parameters are viable, and that the overall algorithm has been implemented correctly. Once this prototype converges, it is a non-trivial problem to improve overall playing strength by extending the training to use a deeper or wider network. On the one hand, while it is simple to retrain completely using the same hyper-parameters, the process of generating the self-play records can be very costly. On the other hand, if we reuse the previously generated self-play records, it is not clear how to initialize the larger network's parameters, nor do we know how the self-play records should be sampled.

Techniques have been proposed to rapidly transfer the information stored in one neural network (NN) (referred to as the parent network) into another NN (referred to as the child network), so that the training process of the larger child network can be accelerated. Net2Net [1] is one such technique that can accelerate training by transferring an NN into another deeper or wider NN without reducing performance in image recognition. Net2Net expands the parent network by adding *identity layers* to it; the output of these identity layers are essentially the same as its inputs, so the child network behaves the same as the parent network initially. While Net2Net has been shown to be useful for CNN architectures, the same technique cannot be easily applied to computer Go, where the building blocks tend to consist of residual networks (ResNets), as in AlphaGo Zero's case [7]. Namely, ResNets contain *shortcut connections* [2] to deal with the degradation problem, which complicates the design of identity layers.

In this paper, we propose a new method to deepen a previously-trained parent ResNet following the AlphaGo Zero algorithm. The expanded child network is able to retain comparable performance, with further potential for training. We then propose two methods to train and further improve this child network, where the training data consists of the same self-play game records. This allows us to skip the most time consuming step in the AlphaGo Zero algorithm. Given the same collection of self-play records, by expanding the 20 block parent network at 3/4 of the overall training progress into 40 blocks, we were able to reach the same level of strength as a randomly initialized 40 block network in only 1/4 of the total training time.

2   Background

In this section, we briefly review residual networks, the AlphaGo Zero algorithm, and the key concepts of Net2Net network extension technique.

2.1   Residual Networks

The ResNet architecture was proposed by He et al. [2] to address the *degradation* problem in DNNs. In short, it is intuitive to assume that deeper networks tend to be better universal function approximators, and so earlier on, researchers have attempted to improve performance by simply increasing the depth of NNs.

Without going into details, two problems can arise from having networks that are simply too deep: the vanishing/exploding gradient problem (solved by techniques such as normalization layers [4]) and the degradation problem. When the degradation problem occurs, performance saturates and converges at a lower accuracy despite having more layers in the NN. Degradation has a different root cause than vanishing gradients, and is not caused by overfitting. By using *shortcut connections* to allow features to skip over one or more layers, as shown in Fig. 1, deeper ResNets are able to overcome the degradation problem and obtain better accuracy than shallower networks.

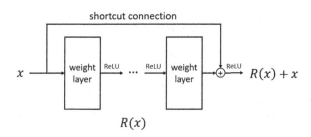

Fig. 1. Illustration of a ResNet block.

2.2   AlphaGo Zero

The goal of AlphaGo Zero is to train a Go agent using no human knowledge except the rules of the game. The algorithm is divided into 3 parts, self-play, optimization, and evaluation. Since this paper focuses on network training, the Monte Carlo tree search (MCTS) component of the AlphaGo Zero algorithm will not be discussed.

First, a randomly initialized network is used by the self-play player initially, denoted by p_0. In each epoch of training, this player continuously plays Go against itself to generate game records until a specified amount of games are collected. We refer to the set of game records generated in one epoch as a *collection*, denoted as c_k, and the self-play player as p_k, where k is the epoch number.

Second, during each epoch, a replay buffer is used to store the r most recent collections of game records, from which the optimization process involves sampling training data from this buffer to optimize the network. The replay buffer does not contain the full repository of game records because earlier collections (say, c_1) tend to be of too low quality for network optimization. In other words, the collections which are in the replay buffer need to match the current network's playing level. The hyperparameter r is referred to as the replay buffer size. During the predefined interval (from a collection newly used to the next), the network weights are saved as checkpoints, whose count is a hyperparameter.

Third, the network at each checkpoint is evaluated according to its win rate against the current self-play player. If the network at some checkpoint wins more than 55% of the games against the current self-play player, the former replaces the latter as the new self-play player. By iterating these three steps, the strength of the self-play player generally improves. During the process the quality of game records will also increase.

2.3   Net2Net

Net2Net is a technique proposed to transfer a smaller network to a larger one [1]. With Net2Net, new layers are added to the original, smaller network (or *parent network*), to form a new, larger network (or *child network*). More specifically, a strategy called the *function-preserving initialization* is proposed, from which the parameter θ' of the child network g can be decided such that

$$\forall x, f(x, \theta) = g(x, \theta') \tag{1}$$

where x is the input data, f is the parent network, and θ is parent network's parameters. The strategy can also be use in partial consecutive networks. As long as strategy is satisfied, the output of the child network g will always be the same as the output of the parent network f. Following this strategy, the methods *Net2WiderNet* and *Net2DeeperNet* were both investigated, where the former tries to widen f and the latter tries to deepen f. Since the scope of this paper focuses on deepening ResNets for the AlphaGo Zero algorithm, we will not discuss Net2WiderNet further.

Net2DeeperNet uses identity layers I to deepen the parent network. Suppose the shape of the identity layer is (C_{in}, C_{out}, K, K) where C_{in} is the number of input channels, C_{out} is the number of output channels, and K is the kernel size, which is usually an odd number. Since the output must be equal to the input to satisfy the function-preserving strategy, $C_{in} = C_{out}$. The kernel of the identity layer at index (m, n) is then as follows,

$$I(m, n) = \begin{cases} identity\ kernel & m = n \\ zero\ matrix & \text{otherwise} \end{cases} \quad 1 \le m, n \le C_{in} = C_{out}. \tag{2}$$

The identity layer can be added anywhere in a network. However, we must take into consideration the activation functions. The network usually consists of an activation function ϕ after a convolution layer. To satisfy the function-preserving strategy, the activation function composition must satisfy

$$\forall v, \phi(I\phi(v)) = \phi(v) \tag{3}$$

where v is a vector. As an example, if ϕ is the sigmoid function, the condition would not be satisfied. On the other hand, a ReLU would be acceptable. As long as the strategy is satisfied, we can deepen a network by adding identity layers at any depth within the parent network.

3   Our Method

In this section, we describe our transfer method and how we train the child network using the parent network's collection of self-play records.

3.1   Transfer Method

We divide the method into two parts. First, we describe the *extension type*, which defines how we add new residual blocks. Second, we describe the *connection type*, which refers to where the new residual blocks are placed.

Extension Type. We introduction three extension types based on the strategy given in Eq. 1. For all three types, suppose that a block B is represented by a function $y = B(x, \theta)$, where y is the output. To add a new block B' after B, we must satisfy the following:

$$\forall x, B(x, \theta) = B'(B(x, \theta), \theta'), \tag{4}$$

where θ' is the parameter of the new block B'. We list the three extension types as follows.

1. Unit-extension: We add two identity layers in the new block. However, due to the shortcut connection architecture of ResNets, we must add a 1/2 scale operator at the end of the new block to ensure the sum remains at a similar scale.
2. Zero-extension: We add two convolutions in the new block. Instead of identity layers, we initialize the weights of the convolutions to be zeros, referred to as *zero layers*. There is no need to add the scale operator. However, this extension might be more difficult to converge during training.
3. Intra-extension: In Net2Net, several identity layers were added in every block. Therefore, we increase two identity layers within the original blocks. That is, with this extension, the number of blocks does not increase.

The activation function used in ResNets tend to be the ReLU function, so by definition Eq. 3 is satisfied. Furthermore, a small noise signal is added so that subsequent training of the child network will not remain at the local optimum of the parent network, and therefore lead to faster convergence. With this addition of noise, the performance of the child network is expected to be slightly worse than the parent network, but we believe that the child network should reach the parent network's performance rapidly, and eventually exceed it. This also ensures that the child network can learn to use the new blocks' capacity.

Connection Type. For clarity of communication, we introduce an encoding system that can represent where the new blocks are added into the parent networks in this paper. We use '1' to represent ten new blocks, and '0' to represent ten original blocks. Ext-ITL represents the new blocks and original blocks interleaved with each other. The encoding string from left to right refers to the child network architecture from the first layer to the last layer. Note that since the Intra-extension method does not create any new blocks, this encoding system does not apply to the method. To illustrate the various connection types, we extend the AlphaGo Zero training from 20 blocks to 40 blocks as an example, shown in Fig. 3.

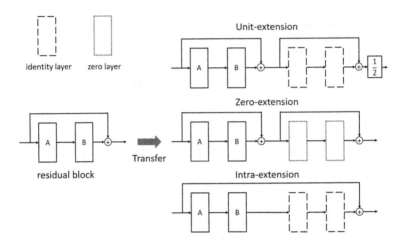

Fig. 2. Illustration of three extension types.

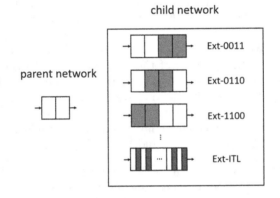

Fig. 3. Connection types. The white squares represent 10 original blocks, and the gray squares represent 10 new blocks. The Ext-ITL connection type represents 20 original blocks and 20 new blocks interleaved with each other.

3.2 Training Method

There are different approaches to sampling the game records when training the child network. More specifically, game records collections from earlier epochs tend to be less suitable for networks at later epochs, and vice-versa. Since the child network is expected to retain a similar level of strength as the parent network, once the parent network is transferred at some epoch, the child network should continue training with the game records from the replay buffer at that epoch. The problem is therefore at which epoch the transfer should take place. We now propose two different methods that describe when the parent network will be transferred, and how the child network should be subsequently trained.

End-Training. For end-training, the transfer occurs for the last self-play player p_n, that is, the last epoch is n. The r most recent collections (i.e. c_{n-r+1} to c_n) are loaded into the replay buffer, and the replay buffer does not need to be changed for the subsequent training of the child network. As shown in Algorithm 1, we transfer the player p_n to the child network. Then we load the last r collections into the replay buffer. Next, we simply train the child network using the replay buffer game records iteratively until a preset maximum number of iterations is reached. One of the caveats of end-training is that if the preset maximum number of iterations is small, the training may not be sufficient; on the other hand, if it is large, the same game records in the replay buffer will be trained many times, possibly leading to overfitting. Thus, another training method is proposed below.

Algorithm 1. End-Training

---

load game records in collections $c_{n-r+1} \sim c_n$;
while *maxIter is not reach* **do**
| select $c_{n-r+1} \sim c_n$ games to train the child network;
end

Shift-Training. When compared to the previous method, the parent network is transferred at epoch i. The game record collections matching the player p_i's level is loaded into the replay buffer. We then iteratively train the child network using the replay buffer, shifting the contents of the buffer to load more recent collections accordingly. As shown in Algorithm 2, we transfer the parent network for p_i to its child network, where i is referred to as the *transfer epoch*. Then we load at most r collections, specifically c_{i-r+1} to c_i, into the replay buffer. Next, we repeatedly load the next collection and train the child network using the replay buffer until the last collection c_n is loaded.

When training the child network, we increase the number of times t a sampled game is used to update the network. That is, since the child network is deeper than the parent network, our expectation is that more back-propagations need to be performed. For this reason, in this paper, each sampled game in shift-training is trained five times (i.e. $t = 5$), unless otherwise mentioned.

Algorithm 2. Shift-Training

---

if $i - r + 1 > 0$ **then**

    load game records in collections $c_{i-r+1} \sim c_{i-1}$;

    head $= i - r + 1$;

else

    load game records in collections $c_1 \sim c_{i-1}$;

    head $= 1$;

end

end $= i$;

while $end \leq n$ **do**

    load game records in collection c_{end};

    if end - $head = r$ **then**

        delete the c_{head} game records from replay buffer;

        head $= head + 1$;

    end

    select $c_{head} \sim c_{end}$ games to train the child network t times;

    end $= end + 1$;

end

---

4   Experiments

We demonstrate our method on 9×9 Go. We follow the AlphaGo Zero algorithm to train a Go agent with 20 ResNet blocks (and with 256 channels), at the end of which we obtain a player p_n. Our goal is to transfer the network to a deeper child network by adding 20 new blocks (consisting of 40 new convolution layers), following up with child network training, and evaluating the resulting player with p_n.

4.1   Experimental Setup

In the following experiments, we evaluate the strength of the resulting child network players by its win rate against the baseline p_n, where both players use 2 s of simulation time with a single NVIDIA Tesla V100 GPU. Since the child network is deeper than the parent network, the forward-pass of the child network takes more time. Nonetheless, the total simulation time is equally set to 2 s for both the child and the parent network. Each transfer method is expressed in terms of its extension type and its connection type. Furthermore, to speed up the overall experiments, for each match up, we continue playing against the baseline until a confidence interval of 95 % is reached to evaluate the probability of having a higher win rate against the baseline than 50%. We show the experiment setup in Table 1.

4.2   Experiment for Shift-Training

We use two experiments to analyze the effect of different transfer methods and transfer epoch i.

Table 1. Experiment hyperparameters and details.

Setting	
Replay buffer size r	20
Collection size	10000
Number of epochs n	243
Total batch size	2048
Learning rate	0.005
Weight decay	0.0001
SGD momentum	0.9
Local memory	754 GB
Thread count	36
Training hardware	8 GPUs (V100)

Comparison Between Transfer Methods. In this experiment, $i = 3n/4$. In addition to the transfer methods, we also trained a control group where the parent network is replaced by a randomly initialized 40 block network at $i = 3n/4$. This control group signifies what happens when no transfer occurs. Table 2 shows the highest win rate of various transfer methods. Unit-extension Ext-0011 has the best performance. It seems likely that the new blocks' capacity enhances the network to recognize more high level features. On the contrary, Zero-extension Ext-1100 performs the worse; it is possible that by introducing new layers, the low level features learned by the parent network were not carried over into the child network. With the exception of Zero-extension Ext-1100, all

Table 2. Result of various transfer methods.

Extension-type	Connection-type	Highest win rate of network (2 s)
Unit-extension	Ext-0011	61.69% ($\pm$4.00%)
	Ext-0110	55.45% ($\pm$3.88%)
	Ext-1100	57.07% ($\pm$3.29%)
	Ext-0101	47.85% ($\pm$3.78%)
	Ext-ITL	53.76% ($\pm$3.83%)
Zero-extension	Ext-0011	45.61% ($\pm$4.09%)
	Ext-0110	52.47% ($\pm$3.79%)
	Ext-1100	23.53% ($\pm$3.88%)
	Ext-0101	56.70% ($\pm$3.90%)
	Ext-ITL	48.15% ($\pm$3.77%)
Intra-extension	X	52.16% ($\pm$3.78%)
Random initialization (control)		42.60% ($\pm$4.34%)

other transfer methods exceed the random initialized control group, showing that the transfer methods can be used to preserve information learned by the parent network, with potential for further growth.

Comparison Between Transfer Epochs. Next, we set the transfer method to be Unit-extension Ext-0011. The transfer epoch i is set to $\{2n/3, 3n/4, 4n/5, 7n/8\}$, as shown in Table 3. The best performing transfer epoch is shown to be $3n/4$. Nonetheless, the win rates for all settings are higher than 50%, indicating that the resulting players can achieve at least p_n level. Earlier transfer epochs i imply more time spent training the more expensive child network, since the child network with shift-training uses 5 updates per sampled game. Therefore, there is a trade-off between transferring earlier (therefore spending more time training) and performance.

Table 3. Result of different transfer epochs.

i	Highest win rate of network (2 s)
$2n/3$	57.00% ($\pm$3.92%)
$3n/4$	61.69% ($\pm$4.00%)
$4n/5$	54.50% ($\pm$3.85%)
$7n/8$	51.92% ($\pm$3.77%)

Evaluation of Shift-Training. According to the previous experiments, Unit-extension Ext-0011 with $i = 3n/4$ is used as the setting for this experiment. We compare with the randomly initialized 40 block ResNet with $i = 1$ and $t = 5$, which is the naive method of retraining the 40 block network from scratch, with all other hyperparameters set equally. We refer to this group as the *40 block retrain* set. Compared with the retrained model, our method requires only about 1/4 of the training time, with its highest win rate to be as high as the retrained model, as shown in Fig. 4.

Additionally, we also retrained a separate 40 block ResNet with the same settings, except $t = 1.3$, so that the total number of training iterations are close to the transferred method. This second model is referred to as the *iteration normalized retrain* set. Our transfer method (Unit-extension Ext-0011) played against the 40 block retrain, iteration normalized retrain, and other high level players, where the result is shown in Table 4. The results show that Unit-extension Ext-0011 can match the 40 block retrain in performance, while exceeding the iteration normalized retrain model significantly. Furthermore, the win rates between Unit-extension Ext-0011 and other high level players are all higher than 50%. This shows that Unit-extension Ext-0011 does not exhibit signs of overfitting.

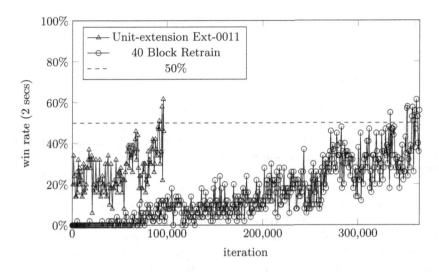

Fig. 4. The win rate curve of our method and the 40 block retrained model.

Table 4. Comparison with other high level players.

Player	Opponent	Win rate (2 s)
Unit-extension Ext-0011	40 block retrain	50.32% (±3.91%)
	Iteration normalized retrain	63.50% (±6.69%)
	p_{232}	89.00% (±4.35%)
	p_{235}	77.00% (±5.85%)
	p_{236}	72.50% (±6.20%)
	p_{238}	67.50% (±6.51%)
	p_{243}	61.69% (±4.00%)

Table 5. Results for end-training.

Extension-type	Connection-type	Highest win rate of network (2 s)
Unit-extension	Ext-0011	39.20% (±4.28%)
Intra-extension	×	64.76% (±4.03%)

4.3 Experiment for End-Training

In this experiment, we analyze two transfer methods of end-training. With end-training, the transfer epoch i can be thought of as fixed at n. As shown in Table 5, the results for Intra-extension seem to be very strong. This could be because the training data (from the replay buffer) is fixed, so it is relatively easier to fit. Playing between Intra-extension with end-training and the 40 block retrain with

2 s simulation time yields a win rate of 42.85%, so it is possible end-training can lead to overfitting.

5 Conclusion

We present a study on extending the ResNet of an agent trained using the AlphaGo Zero algorithm. We propose a scheme which transfers a parent network to a deeper child network without losing the learned knowledge; additionally, we propose two methods of sampling game records generated by the parent network for training the child network. Overall, the child network can retain the parent network's performance, or even surpass it with further training. From our experiments, we analyze a collection of hyperparameters to conclude that the Unit-extension Ext-0011 transfer method performs the best, with a win rate of 61.69% against the strongest player from the parent network. Finally, We demonstrate our method only requires about 1/4 of the training time as completely retraining a network of the same size, while simultaneously achieving the same level of performance.

Further research is left open to investigate more general extension-types in the future which can then be used for various network architectures, such as VGGNet [8], AlexNet [5], and DenseNet [3].

References

1. Chen, T., Goodfellow, I., Shlens, J.: Net2Net: accelerating learning via knowledge transfer. arXiv preprint arXiv:1511.05641 (2015)
2. He, K., Zhang, X., Ren, S., Sun, J.: Deep residual learning for image recognition. In: Proceedings of the IEEE Conference on Computer Vision and Pattern Recognition, pp. 770–778 (2016)
3. Huang, G., Liu, Z., Van Der Maaten, L., Weinberger, K.Q.: Densely connected convolutional networks. In: Proceedings of the IEEE Conference on Computer Vision and Pattern Recognition, pp. 4700–4708 (2017)
4. Ioffe, S., Szegedy, C.: Batch normalization: accelerating deep network training by reducing internal covariate shift. arXiv preprint arXiv:1502.03167 (2015)
5. Krizhevsky, A., Sutskever, I., Hinton, G.E.: ImageNet classification with deep convolutional neural networks. In: Advances in Neural Information Processing Systems, pp. 1097–1105 (2012)
6. Pascutto, G.C.: Leela-Zero Github repository (2018). https://github.com/gcp/leela-zero
7. Silver, D., et al.: Mastering the game of Go without human knowledge. Nature 550(7676), 354 (2017)
8. Simonyan, K., Zisserman, A.: Very deep convolutional networks for large-scale image recognition. arXiv preprint arXiv:1409.1556 (2014)
9. Tian, Y., et al.: ELF OpenGo: an analysis and open reimplementation of AlphaZero. arXiv preprint arXiv:1902.04522 (2019)
10. Wu, I.C., Wu, T.R., Liu, A.J., Guei, H., Wei, T.: On strength adjustment for MCTS-based programs. In: Thirty-Third AAAI Conference on Artificial Intelligence (2019)

Designing Policy Network with Deep Learning in Turn-Based Strategy Games

Tomihiro Kimura$^{(\boxtimes)}$ and Kokolo Ikeda

Japan Advanced Institute of Science and Technology, JAIST, Ishikawa, Japan
kt499887@gmail.com, kokolo@jaist.ac.jp

Abstract. Research on artificial intelligence (AI) has experienced a substantial stride since the advent of the AlphaGo, progressing the application of deep learning techniques for the game application. However, significant research is still unpublished in the field of turn-based strategy games, owing to the complexity of the game structure and its computational problem. To apply deep learning to turn-based strategy games, a policy network created from match data was developed from learning game records. The neural network design used as a policy network is integrated into the turn-based strategy games, using a recurrent neural network to reduce the number of output neurons and to divide the output structure into original positions, destinations, and attack positions. Using the state and action data as a database, the game data are generated from the learning map based on the competition with the Monte Carlo Tree Search (MCTS) algorithm. However, the produced policy network demonstrates a superior performance against the MCTS algorithm with a winning rate of over 50% on the learning maps, and over 40% on the validation maps. In the game, the thinking time for the deep learning is extremely short since this it is performed by inference only, whereas MCTS thinking the time is approximately 5 to 10 s per move.

Keywords: Turn-based strategy game · Deep learning · Recurrent neural network · Policy network

1 Introduction

Research on game of artificial intelligence (AI) agent has recorded great breakthroughs since the advent of AlphaGo, that combines highly sophisticated deep learning and reinforcement learning techniques. AlphaGo has advanced to the level of beating professional Go players, demonstrating high performance in the game [1–3]. In many game domains, AI players have improved more tremendously than humans. However, humans have surpassed the AI agents in the game of turn-based strategy because AI agents still possess inadequate strength levels, and fewer scientific researches have just been published in this area.

In the present study, we describe difficulties in the game of turn-based strategy to accelerate its research framework. Then, we demonstrate a method of

© Springer Nature Switzerland AG 2020
T. Cazenave et al. (Eds.): ACG 2019, LNCS 12516, pp. 143–154, 2020.
https://doi.org/10.1007/978-3-030-65883-0_12

solving the data complexity problem by applying recurrent networks success-fully. Furthermore, we apply deep learning to avoid a computational explosion. Finally, we build and evaluate a policy network at a level turn-based strategy games are played. Typically, we apply deep learning to design the game policy network and to integrate it with a Monte Carlo Tree Search (MCTS) algorithm or value network. Traditionally, policy networks have been learned from game records, however, learning on a map is still difficult because of the application of complex data structure of a game to deep learning.

2 Turn-Based Strategy Games as an Academic Platform: TUBSTAP

2.1 Difficulty of Turn-Based Strategy Games

The difficulty of turn-based strategy games is based on the following features:

1. Huge search space beyond Shogi and Go.
2. Initial board condition is unfixed.
3. Various characteristics are elements of units.
4. Several types of terrain.
5. Multiple units operate in one turn.

While a large amount of previous game record data is accumulated in chess, Shogi, and Go, this is unavailable in the turn-based strategy game. Thus it is necessary to create a new one.

To promote research that is being delayed, we proposed Turn-Based Strategy Games as an Academic Platform: TUBSTAP [4] despite its long history and pop-ularity in the game market and its lack research platform in turn-based strategy games. In TUBSTAP, it is possible to compare the performance by develop-ing, evaluating and researching a turn-based strategy game AI agent without using additional work. The game rules are examined, extracted, simplified, and abstracted from the rules of various turn-based strategy such as "Daisenryaku" and "Famicon Wars". With this, it is possible to check the basic operation of the turn-based strategy game on this platform [5]. Although microRTS [6] exists, which is a framework similar to TUBSTAP and target real-time strategy games, the difference is its real-time game structure with a relatively large-scale system.

2.2 TUBSTAP Game System

Infantry, Panzer, Cannon, Anti-air Tank, Fighter planes, and Attack Aircraft are units in the TUBSTAP game. Six types of land cells are mountain, forest, plain, road, sea, and fortress. All units have an initial HP(Hit Point indicating the life of unit) integer value between one and ten, removed from the board when it reaches 0. When the units attacked, the HP decreased; simultaneously, the counterattack reduced the attacking side unit HP. In many ways, TUBSTAP is similar to the "Famicom Wars", the game progresses by fighting between the RED and BLUE forces.

The match ends when either units are completely destroyed, or the specified number of turns is reached. The winner is either annihilated all opponent's units or, the winner has a difference in the total number of HP at the end of all turns that exceed the specified value.

TUBSTAP allow users to create and test a wide variety of maps. Typical map examples are shown in Fig. 1.

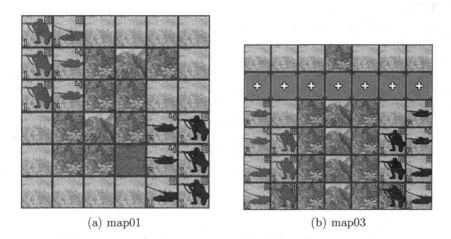

(a) map01   (b) map03

Fig. 1. Map01 is the most basic type of map of 6 × 6 square size. Map03 is a 6 × 7 size map where cannon plays an active role.

2.3   Related Works for TUBSTAP

The algorithms applicable to TUBSTAP are limited owing to the difficulties in turn-based strategy games. MCTS [7] and its variations are well-known algorithms. As another type, a method of searching a tree by dividing the positions of units has been published [8,9]. Other algorithms are still unpublished; thus, we applied deep learning to TUBSTAP [10]. However, the match setup was limited to single units.

Benchmark Maps. We developed and proposed a set of benchmark problems to compensate for the small number of maps and to increase diversity [11]. These benchmark problems are from simple to difficult levels, such as escape tracking problems, pinch problems, pathfinding problems, and multi-unit coordination problems used to evaluate the standard performance of AI.

TUBSTAP AI Tournaments. In recent years, TUBSTAP AI tournaments have been held in Japan, and several AI agents compete for performance [12,13]. Many of the AI agents who participated in the competition were open and downloadable.

3   Purpose and Procedure

The purpose of this experiment is to propose a neural network design suitable for a turn-based strategy game using a complicated data structure.

In this study, the experiment consists of three phases: the first is the creation of game record data. The second is the learning of the neural network, and the third is the verification of the match, as shown in Fig. 2. The game is a battle between infantry on a plain with no obstacle, and the infantries are two: each in RED and BLUE. Using a standardized neural network, the overall map size was fixed at eight squares; however, the battlefield was reduced to five squares for easier fighting.

We expects that deep learning distills and extracts tactics and strategies from a database and transfers them into a neural network. To check the generalization capabilities of deep learning, learning validation is performed on wider maps, not on the map for learning. Moreover, it checks the learning capability of the designed neural network by competing with other algorithms on an unknown map.

Learning Maps. The map for learning is a battlefield of 5 cell × 5 cell square-shaped plains surrounded by no-entry cells, as shown in Fig. 2.

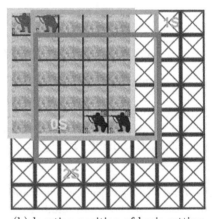

(a) sq5x5 basic setting   (b) location position of basic setting

Fig. 2. Learning maps. The boundaries are shifted by one cell per map like 0s, 1s, and 2s, to avoid bias and ensure diversity.

3.1   Previous Studies

For the last decades, neural networks have been applied to the field of artificial intelligence, using many approaches to learn game records. Neurogammon [14], adopted the multilayer neural network, learning from the expert dataset to win a Gold medal in the 1st Computer Olympiad 1989. In the domain of Go, Deep Convolutional Neural Network [15], learned from a human's professional database demonstrated a high winning rate without search results, and AlphaGo's policy network achieved better performance later [1].

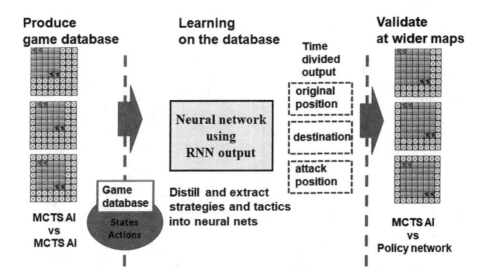

Fig. 3. Experimental procedure

4   Experiments

4.1   Game Record Database for Learning

The game records for learning are created under the following two policies. One uses AI agent running on TUBSTAP to create a match record from the initial state of the learning map. The second uses the same AI agent to create a set of additional match record from various situations which some specific tactics are needed to win. This is a tactical procedure.

The sizes of this database are approximately 800,000 in the number of game phases and approximately 100,000 number of games. When the game is over, only the winner's status and actions are incorporated into the database to enhance the superior experience.

The M-UCT is the AI agent, a top performer in the 2016 Game AI Tournaments in Japan [12], M3Lee won the 2nd place in the same competition. And the tree search algorithm based on the self-made alpha beta search is also used. M-UCT runs 5000 simulations and rolls out to the terminal. And M3Lee runs 3000 simulations.

Figure 2 shows the basic setting of learning maps. There are only two infantry units on a simple terrain, so it may look quite easy. However, even with this settings, a bit difficult or unexpected tactics are sometimes needed to win (we will show examples later). Then we chose these maps as the first step.

Validation Maps. After learning the neural network, the maps that verify the generalization performance of the network are validation maps, as shown in Fig. 4. Validation maps are not used for learning.

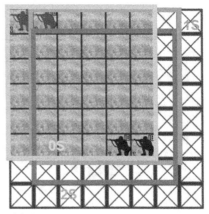

(a) sq6x6 basic setting   (b) location position of basic setting

Fig. 4. Validation maps. These maps are used only for verification tests. The boundaries are shifted by one cell per map like 0s, 1s, and 2s, to avoid bias and ensure diversity.

Tactical Procedures. Figure 5 shows part of an example of the tactical procedures for the layout map. In the initial phase, attacks as indicated by the arrow effectively reduces the opponent's HP and increases the chance of winning the game. We considered more than 30 patterns of such tactical procedures, and prepared over 400,000 data by hands.

Data Augmentation. To enrich data and prevent bias in the neural network, rotation, and mirror data augmentation are performed. There are four times as large as rotation (0/90/180/270-degrees) and twice as large as a reflection (vertical/horizontal), eight times of data in total.

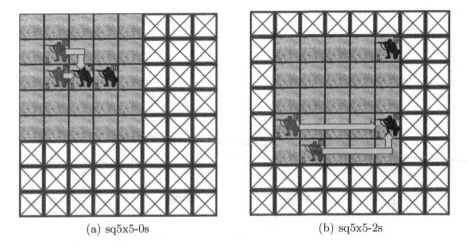

(a) sq5x5-0s   (b) sq5x5-2s

Fig. 5. Tactical procedure examples. The arrow shows the attack target to rob the opponent's HP effectively.

4.2 Neural Network Design

Designing neural networks for turn-based strategy games is complicated and difficult owing to many features such as terrain features and unit features. To use a neural network as a policy network, an action unit instruction unit must be output. Moreover, the complicated data output is essential, and we propose a high performance design suitable for turn-based strategy games.

Input Data Structure. All input channels to the neural network are encoded into 7 layers of terrain data, 14 layers of unit data(the type of units and moved record for each RED/BLUE), and 1 layer of turn data.

Output Data Structure. The minimum information to describe the unit's behavior in TUBSTAP is the unit's original position, destination, and attacking unit position. Although the neural network must output these signals, the conventional one-hot output representation requires an enormous number of neurons. In the case of a map of 8×8 squares, at least $64 \times 64 \times 64 = 262144$ neurons have been required since $8 \times 8 = 64$ neurons are multiplied by the original position, destination and attacking position. A design with only this neuron is unrealistic. Therefore, we aim to reduce the number of output neurons and to simplify the design by the time-dividing the output using a recurrent neural network (RNN). As shown in Fig 6, the output structure of the designed neural network is illustrated in its schematic diagram (unfolded in time). The output of $t = 1$, indicating the original position, is connected to the input $t = 2$. The output of $t = 2$, indicates the destination and connected to the input of $t = 3$. This connection enhances correlation and reduces inference errors. Usually,

RNN is effective for the output of the time division. Thus, the GRU unit [16] was adopted instead of the LSTM unit in consideration of the execution speed. The GRU unit is reset at the end of t = 3. The program corrects the output by legal move masks.

The neural network outputs the behavior of one unit at a time. In order to define the behavior of two units, it is necessary to perform two inferences.

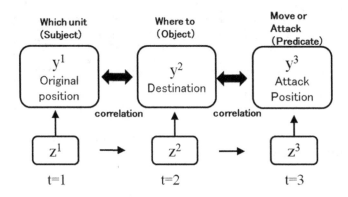

Fig. 6. Output data structure

Neural Network Structure. The configuration of the neural network is as follows: As shown in Fig. 7, a three-layer Conv2D network is in the input state. Three affine layers, are in the next stage, fully connected layer; connected with batch normalization and L2 regularization In the next stage, a further recurrent network for time division output is connected with Dropout and ReLU. In the affine and recurrent layers, 800 neurons were used for each layer.

4.3   Learning Procedure

Since the game record file size is huge and cannot load entire files into the PC memory, a buffer memory is prepared, sampled randomly, and used for learning of the neural network. In this case, the buffer memory acts like a replay memory in reinforcement learning [17]. The buffer memory size is 12000 game phases, and the data of one file is updated for each iteration. In this learning procedure, it is 1 epoch per iteration, the batch sizes are 2048, executed up to 7000 iterations. The optimizer used Adam, and loss function was set to cross entropy type. Neural networks start learning from zero, the initial state of random weights. Therefore, no need for other data or settings other than the database used.

In the game match process, several correct procedures are available not limited to only one answer in many cases.

Therefore, common measurements such as accuracy or recall used in usually supervised learning researches are not sufficient to evaluate the performance of

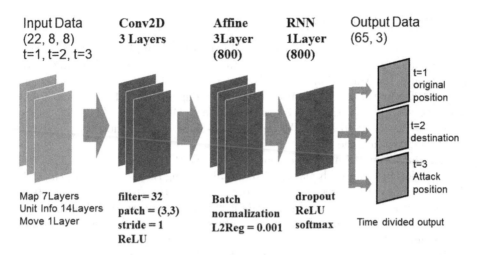

Fig. 7. Neural network structure

trained a policy network. In our case, the accuracy reached 59.5% after 7000 iterations. We can compare such measurements of different methods (such as different network structure), but we cannot say this policy network is good or not in an absolute manner. Thus, we evaluate the performance through the actual matching against MCTS player.

Policy Network Accuracy. Table 1 shows the data of the matching rate between the 10,000 phases extracted from the produced database and the output of the learned neural network.

Table 1. Matching rate between the 10,000 phases extracted from the database and the policy network output

Iteration	Accuracy
1000	29.8%
5000	37.6%
7000	59.6%

Learning Rate. Learning rate scheduling was performed to speed up the learning process at 10^{-3} up to 1000 times, 10^{-4} up to 5000 times, and 5×10^{-5} up to 7000 times.

Hardware and Software. In this experiment, Python 3.6 and Keras 2.2 were the software. The CPU Intel i7 3.4 GHz and GPU NVIDIA GTX 1050 Ti were the hardware specifications.

4.4   Result of the Competition Experiment

To evaluate the performance of the neural network to complete learning, a total of 300 games, 100 games for each map, were carried out on the learning maps (Fig. 2) and the validation maps (Fig. 4), respectively. Of the 100 games, 50 played RED first, and BLUE moved the remaining 50 games first. As an opponent, we set M-UCT that operates on standard MCTS. The AI agent in the under test assigned to RED. Without a look-ahead search, the policy network assistes in deciding the action.

On the Learning map, the match result is Win/Lose/Draw = 55/45/0, 53/42/5, 60/40/0, for each map, and the total winning rate is 56.0%. On the validation map, the match result is Win/Lose/Draw = 50/42/8, 35/54/11, 37/59/4, for each map, and the total winning rate is 40.7%. This degraded is approximate 10% compared with the learning map, leading to an increased draw. The results are as shown in Fig. 8.

In the battle, the output of the neural net showed that the scenes of several independent games were integrated and beyond the MCTS algorithm.

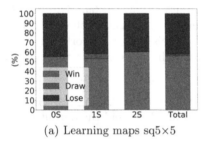 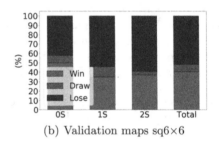

(a) Learning maps sq5×5   (b) Validation maps sq6×6

Fig. 8. Results of the match against M-UCT. 100 matches per map. 0S indicates sq5x5-0s map, similarly 1S/2S indicates sq5x5-1s and sq5x5-2s map respectively.

5   Discussion and Conclusion

The results of the learning reached a level in which the winning rate exceeded 50% against the MCTS algorithm of the learning map; starting from zero the initial state of random weights. In the validation map, the performance exceeded 40%; even though only the action decision of the policy network was used. Validation maps, similar in shape to learning maps, are larger in size and are unused for learning. Therefore, the learning ability of the neural network designed in this study produced high performance. In the game, the thinking time on the deep learning side is extremely short as it is performed only by inference, whereas MCTS side thinking time is about 5 to 10 s per move.

We introduce a designing policy network using the refinement of the recurrent network in turn-based strategy game when the application of deep learning is difficult. The produced policy network demonstrates a superior performance on the unknown map against the MCTS algorithm without a look-ahead search.

We know the current work is a bit premature, for example, different network structures with/without using RNN are not compared, tree search is not combined with the trained policy network, etc. So, we will try to solve such problems soon, especially we intend to use more complex maps such as many and various units are used, and diverse terrains are employed.

References

1. Silver, D., et al.: Mastering the game of Go with deep neural networks and tree search. Nature **529**(7587), 484–489 (2016)
2. Silver, D., et al.: Mastering the game of Go without human knowledge. Nature **550**(7674), 354–359 (2017)
3. Silver, D., et al.: A general reinforcement learning algorithm that masters chess, shogi, and Go through self-play. Science **362**(6419), 1140–1144 (2018)
4. Fujiki, T., Ikeda, K., Viennot, S.: A platform for turn-based strategy games, with a comparison of Monte-Carlo algorithms. In: IEEE Conference on Computational Intelligence and Games, CIG2015, pp. 407–414 (2015)
5. TUBSTAP Homepage. http://www.jaist.ac.jp/is/labs/ikeda-lab/tbs_eng/index.htm. Accessed 10 Apr 2019
6. Stanescu, M., Barriga, N.A., Hess, A., et al.: Evaluating real-time strategy game states using convolutional neural networks. In: IEEE Computational Intelligence and Games (CIG) (2016)
7. Kato, C., Miwa, M., Tsuruoka, Y., Chikayama, T.: UCT and its enhancement for tactical decisions in turn-based strategy games. In: Game Programming Workshop 2013, pp. 138–145 (2013). (in Japanese)
8. Sato, N., Ikeda, K.: Three types of forward pruning techniques to apply alpha beta algorithm to turn-based strategy game. In: IEEE Conference on Computational Intelligence and Games, CIG2016, pp. 294–301 (2016)
9. Sato, N., Fujiki, T., Ikeda, K.: An approach to evaluate turn-based strategy game positions with offline tree searches in simplified games. In: Game Programming Workshop 2015, pp. 61–68 (2015). (in Japanese)
10. Kimura, T.: Simple data representation method with deep learning for turn-based strategy game. IPSJ SIG technical reports, 2019-GI-41, vol. 5, pp. 1–8 (2019). (in Japanese)
11. Kimura, T., Ikeda, K.: Offering new benchmark maps for turn based strategy game. In: Game Programming Workshop 2016, pp. 36–43 (2016). (in Japanese)
12. Game AI Tournaments Homepage (GAT). http://minerva.cs.uec.ac.jp/gat_uec/wiki.cgi?page=FrontPage. Accessed 10 Apr 2019. (in Japanese)
13. Game Programming Workshop Homepage (GPW). http://www.ipsj.or.jp/sig/gi/gpw/index-e.html. Accessed 10 Apr 2019
14. Tesauro, G.: Neurogammon wins computer Olympiad. Neural Comput. **1**(3), 321–323 (1989)
15. Clark, C., Storkey, A.: Training deep convolutional neural networks to play Go. In: ICML 2015 Proceedings of the 32nd International Conference on International Conference on Machine Learning, vol. 37, pp. 1766–1774. JMLR (2015)

16. Cho, K., et al.: Learning phrase representations using RNN encoder-decoder for statistical machine translation. In: Proceedings of the 2014 Conference on Empirical Methods in Natural Language Processing, pp. 1724–1734 (2014)
17. Mnih, V., et al.: Human-level control through deep reinforcement learning. Nature **518**(7540), 529–533 (2015)

On Strongly Solving Chinese Checkers

Nathan R. Sturtevant[(⊠)]

University of Alberta, Edmonton, AB, Canada
nathanst@ualberta.ca

Abstract. Chinese Checkers is a game for 2–6 players that has been used as a testbed for game AI in the past. The game is easily scalable to different size boards, different numbers of players, and different numbers of pieces for each player. In this paper we provide an overview of what is required to strongly solve versions of the game, including a complete set of rules needed to solve the game. We provide results on smaller boards with result showing that these games are all a first-player win.

Keywords: Solve · Game · Chinese Checkers

1 Introduction

This paper studies the problem of strongly solving variants of Chinese Checkers, and provides the foundation needed to strongly solve them. Chinese Checkers is closely related to the game of Halma, which was invented around 1883–1884, except that Chinese Checkers is played on a star-shaped board while Halma is played on a square board [4]. It is typically played by 2–6 players, with the goal of moving your pieces across the board into a goal area before your opponent. There are two reasons why we are interested in studying this game and strongly solving variants of the game.

First, many traditional two-player perfect information games, such as Connect-Four [2], Awari [9], Checkers [13], and Hex [6] have been strongly solved or weakly solved, so solving Chinese Checkers follows in this line of work. One common board size for Chinese Checkers has a 1.73×10^{24} states, and thus might be weakly solvable, given a good proof strategy. However, in our work we have found it difficult to construct small proofs for the game. Thus, by strongly solving small versions of the game, we can study the nature of the game and learn how to build compact proofs for the game.

Second, recent work on AlphaZero [14] has suggested a common approach for learning to play deterministic two-player games with perfect information. As Chinese Checkers falls into this category, we expect that the AlphaZero approach is able to learn to play the game. But, given that we can strongly solve some board sizes, this offers the opportunity to precisely measure and evaluate how learning takes place in the game. Thus, we suggest that Chinese Checkers will be a good testbed for evaluating learning in a two-player deterministic perfect information game.

© Springer Nature Switzerland AG 2020
T. Cazenave et al. (Eds.): ACG 2019, LNCS 12516, pp. 155–166, 2020.
https://doi.org/10.1007/978-3-030-65883-0_13

Given this reasoning, we have built a number of solvers for the game of Chinese Checkers with different board sizes and different numbers of pieces on the board. Through analysis of these solvers we can now provide a deeper analysis of Chinese Checkers than has previously been found in the literature, including comprehensive rules for wins, losses, draws, and illegal positions in the game. This paper provides an overview of these insights, as well as the results of strongly solving games with up to the 6 × 6 board with 6 pieces per player which has 2,313,100,389,600 positions. All games that we have solved have been a first-player win.

2   Background and Related Work

Victor Allis defined three different types of solved games. These are defined as [1]:

- **Ultra-weakly solved.** For the initial position(s), the game-theoretical value has been determined.
- **Weakly solved.** For the initial position(s), a strategy has been determined to obtain at least the game-theoretical value of the game, for both players, under reasonable resources.
- **Strongly solved.** For all legal positions, a strategy has been determined to obtain the game-theoretic value of the position, for both players, under reasonable resources.

While reasonable resources might change over time, it is suggested that a limit of several minutes of computation per move should be allowed. Thus, storing the result of every position would qualify for a game being strongly solved. But, sometimes there are too many positional combinations to store efficiently, requiring the use of search to dynamically re-compute this data when making queries about a state. In such cases, the a game would still be considered to be strongly solved even if it took a few minutes to perform these re-computations.

Given these classifications, significant research has gone into both solving games or and subgames of a game, such as in endgame databases [3,12].

Connect-Four was one of the first non-trivial games to be solved. One of the original proofs used specialized knowledge to build a small proof tree for the game [2], although the full game can now be enumerated efficiently [5], so the game is now strongly solved. Awari [9] was strongly solved using parallel hardware; the value of 889,063,398,406 positions was determined during this proof. Pentago [7] has also been strongly solved on parallel hardware, with 3,009,081,623,421,558 positions in the game. In the Pentago solution positions with more than 18 stones are not recorded on disk, as it requires too much storage and the value of the states can be re-computed easily.

Weakly solved games include Checkers [13], which is solved for the primary opening moves, and 8 × 8 Hex [6], which is solved for any opening move.

Chinese Checkers has been a common domain used in testing for multi-player games [11,16,18]. A common heuristic used for play is the single-agent distance

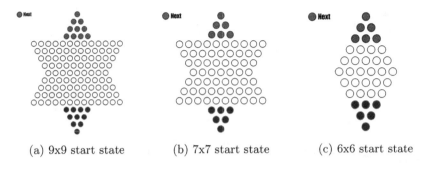

(a) 9x9 start state   (b) 7x7 start state   (c) 6x6 start state

Fig. 1. Examples of different board sizes.

to the goal [16], which can be easily calculated on smaller board sizes. The single-agent distances on the full board size can also be calculated [19] and used for game play [15], but other approaches such as MCTS [10] and TD learning [15] have also been explored for playing Chinese Checkers.

2.1 Terminology

In this paper we use the terms *board*, *board position*, and *state* interchangeably. The term *jumps* can be substituted anytime we use the term *hops*. We refer to pieces on the board, but these can also be referred to as *marbles*, as physical versions of the game often use marbles for pieces. We will also talk about *corners* of the board and goal or start *areas* interchangeably.

3 Rules of Chinese Checkers

If we wish to strongly solve variants of Chinese Checkers, we must be able to determine the value of every state in the game. This requires resolving a number of special cases around how the game is played. In this section we work through and propose resolutions for each of these special cases. This paper focuses primarily on the two-player game.

Chinese Checkers is typically played on a star-shaped board with six corners, as shown in Fig. 1(a) with 10 pieces per player. In the two-player game, player 1 and player 2's pieces start at the top and bottom of the board, respectively, and the goal is for a player to get their pieces into their goal area, which, in this case, is the corner where the opponent's pieces started. Pieces are not typically allowed to move into the other corners of the board, although the rules sometimes allow for pieces to move through these areas as long as they do not remain in one of the other corners at the end of a turn.[1]

---

[1] Our current two-player implementation does not allow this, but we are considering adding this for 7 × 7 boards. The rule does not seem to play an important role in optimal play and may be more useful in the *n*-player version of the game ($n > 2$) when space is more constrained.

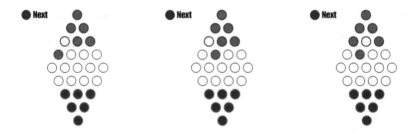

Fig. 2. Adjacent moves in Chinese Checkers.

Given that a player cannot move into corners of the board besides their start and goal corners, the primary game play takes place on the center diamond of the board, which can be represented as a $m \times m$ grid. The game in Fig. 1(a) has a 9×9 gameplay area with 10 pieces per player. Another common board size used in research [16] is shown in Fig. 1(b), which is a 7×7 gameplay area with 6 pieces per player. It is possible to draw the board as a star when the size of the gameplay area is odd. If we use an even-sized gameplay area, then it is not possible to draw uniform-sized corners, and thus it is only to play these sized boards with two players, as shown in Fig. 1(c). In the remainder of the paper we will only draw the main gameplay area, and we will give each player anywhere from one to six pieces.

3.1 Movement Rules

In the game of Chinese Checkers there are two types of moves that can be performed, moves to an adjacent location, or hops over adjacent pieces to non-adjacent locations. Figure 2 shows moves to adjacent locations from the starting position of the game. There are six possible adjacent moves at the beginning of the game; three of the moves are shown in the figure. The remaining moves result in positions that are symmetric to the ones shown, something that will be discussed in Sect. 3.5.

The second type of move hops over an adjacent piece to land on an empty location on the other side. A hop from the starting position is illustrated in Fig. 3(a). If multiple hops are available for a single piece, they can be chained together into longer hops that cross the entire board. Figure 3(b) sets up an alternate board position where additional hops can be chained, allowing a piece to move across the board, giving the position in Fig. 3. A piece can take any number of legal hops in a turn, and can stop when further hops are still possible. Additionally, pieces are not removed from the board when they are hopped over by another pieces.

One notable feature of Chinese Checkers is that the game is not acyclic – it is possible to return to the same position and not make progress. This makes Chinese Checkers more complex to analyze than games like Pentago, which adds a new piece to the board every turn.

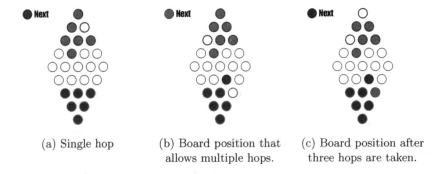

(a) Single hop

(b) Board position that allows multiple hops.

(c) Board position after three hops are taken.

Fig. 3. Hopping moves in Chinese Checkers.

3.2 Winning Conditions

In the most general case a player wins a game of Chinese Checkers if they get all of their pieces into their goal area on the opposite side of the board. In normal play this is achievable, but for exhaustive analysis we need a more precise definition of the winning conditions. In particular, it has been observed that a single player can leave one piece in their start area, preventing the opponent from ever getting their pieces into the goal. This rule is typically not handled in rule books for the game, although web sites discussing the game often make suggestions for handling this. One suggested handling is as follows:[2]

> "If one or more of a player's marbles are imprisoned in his/her original starting point so that the marbles cannot be moved, such player forfeits the game. In the event of multiple players, the imprisoned marbles are removed and the game continues as before."

However, this rule is inadequate. It is possible for a player to place a piece on the outer row of their goal area so that it cannot be imprisoned, but it will still block the other play from winning. Other suggestions that we have seen for handling this have also been inadequate. Thus, we propose the combination of two rules for wins and illegal states to force an end to the game in these types of situations. The first has been traditional used in our implementation of the game, while the second was adapted based on our ongoing work.

Definition 1. *A state in Chinese Checkers is* won *for player n if player n's goal area is filled with pieces, and at least one of the pieces belongs to player n.*

---

[2] http://www.abstractstrategy.com/chinese-checkers-g.html#c-checkers-rules.

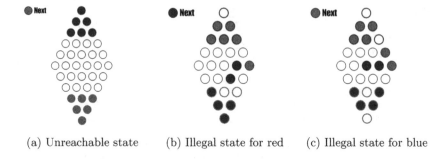

(a) Unreachable state   (b) Illegal state for red   (c) Illegal state for blue

Fig. 4. Illegal states in Chinese Checkers. (Color figure online)

Under this definition if a player leaves some pieces in their goal area, the other player can just fill in around these pieces in order to win the game. For very small versions of the game this can have unintended consequences, as shown in Fig. 3(c). After only 3 ply this state qualifies as a win for the red player, because the red player has a piece in their goal area, and the remainder of the home is filled with blue pieces. Such shallow goal states are not found on larger boards, and the small boards are not used widely, so the shallow goal states seem acceptable. Note that even Chess has possible goal states just 4 ply into the game.

This definition has been adequate for our past experiments with Chinese Checkers playing agents, but it is not adequate for strongly solving the game when there are six pieces on the board, as there are still some conditions it cannot catch. However, instead of modifying our definition of a winning state, we declare certain board positions to be illegal, which we discuss in the next section.

3.3   Illegal States

In Chinese Checkers there are states that are unreachable via normal play. One example of an illegal position is shown in Fig. 4(a). In this state both players have their pieces in the goal area. But, because a player should win as soon as their pieces are moved into the goal area, it is impossible for both players to simultaneously have their pieces in their respective goals. Thus, this state is not reachable by normal play and should be excluded from any analysis that attempts to strongly solve the game. We do this by declaring the state to be illegal. As long as all illegal states are marked or checked appropriately, they can then be ignored in any proof procedure.

Given the definition of a winning state in the previous section, there are a large number of illegal states in the game – all combinations of states where both goals are filled with pieces. Thus, we form a general rule for illegal states, the first of two rules that are required for strongly solving the game.

Definition 2. *(Part 1) A state in Chinese Checkers is* illegal *for player n if the winning conditions are met for player n and it is player n's turn to move.*

This definition implies that it is illegal for a player to take a suicidal action that causes the other player to win. For instance, a player might move back into their own start area to fill it up when the opponent already has a piece in this area, thus creating a win for the other player even when it was their turn to move.

The other type of state that we declare to be illegal are states where a player is blocking the other player from filling their goal area. This type of position is shown in Fig. 4(b). In this state the red player has blocked the outer two rows of their start area. In such a position it is impossible for the blue player to move a single piece into their goal area, even though one of the locations in the goal area is open. This allows the red player to prevent the blue player from winning, because a game is only won when the goal area is filled with pieces, and in this case the red player can always move its other piece instead of unblocking the goal.

While we could declare a state to be illegal if two consecutive rows are blocked with empty locations behind them, our experiments showed that this rule is still inadequate, as shown in Fig. 4(c). This state shows an arrangement of the blue pieces at the bottom of the board that is sufficient to prevent the red player from filling the goal without blocking two consecutive rows. We use this position to formulate the second condition for a state to be illegal.

Definition 3. *(Part 2) A state in Chinese Checkers is illegal if there are one or more unoccupied locations in player n's goal area that are unreachable by player n due to another players pieces.*

These rules are only necessary on game variants with more pieces. When there are six pieces, it suffices to check the two outer edges of the goal area to see if they are occupied, and if the tip of the goal is unoccupied. On larger boards, when a player has 10 pieces, the conditions are more complicated because there are more ways to block locations within the goal area.

Under this rule the states in both Fig. 4(b) and (c) are illegal, meaning it is illegal to take an action that leads to this state. This solves one particular problem; the question remains whether it creates alternate problems in gameplay. In looking at the shortest single-player sequence of actions needed to cross the board, such positions are never reached. This is because players are focused on getting their pieces out of the start area as quickly as possible, not on fortifying the start area.

For human play against suboptimal opponents, these positions could appear as part of normal play, and we do not dismiss the possibility that we could be missing some important cases. But, in human play it is also possible to use reasonable judgement to determine if a player is intending to block the goal, and thus such a rule would only need to be applied selectively. But, since we do not have that luxury in optimal computer vs computer play, we choose to err on the side of simpler rules.

3.4 Draws

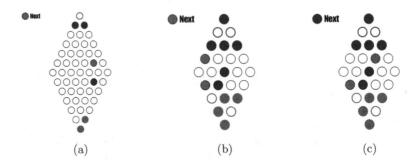

Fig. 5. Drawn states in Chinese Checkers.

In some games, such as Hex, all states can be determined to be a win or loss. The possible value of states in Chinese Checkers has not, to our knowledge, been previously determined. That is, we have unable to find any rules for the game that discuss drawn positions. If we prove all possible states that are wins, losses and illegal, yet have states that are still unable to be proven, these remaining states are drawn. Drawn states begin to appear in Chinese Checkers once there are three pieces on the board.

We illustrate one such state in Fig. 5(a). On this board each player has two pieces in the goal area, and one piece left in the middle of the board. Ignoring the blue piece, the red piece in the middle can reach the goal area in six moves. Similarly, the blue piece in the middle can also reach its goal area in six moves. However, if the red player takes the first move, it will allow the blue player to perform a jump and reach the goal in five moves, thus winning the game. Thus, this similar to a *zugzwang* position, where the player to move is at a disadvantage, except that there are other pieces that can be moved in order to delay the disadvantageous move. Instead of making progress, both player will alternate moving one of their two pieces that have already reached the goal.

Figures 5(b) and 5(c) illustrate a more complicated position that is also drawn. In 5(b) the red player moves to the state shown in 5(c) to block the blue player from performing a double hop into their goal area. As a result, the blue player will move its piece in the middle row one step to the left in order to enable a different double hop. After this the red player will undo its previous move, returning to block the blue piece. This then causes the blue player to move back to the position in 5(b). In this set of positions blue has an advantage, but red can continually block that advantage.

These figures illustrate that drawn positions can occur in Chinese Checkers. As existing rules do not account for repeated positions, we suggest the following rule.

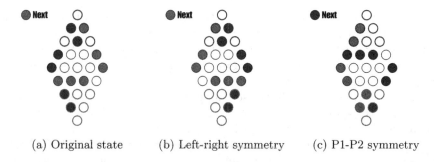

(a) Original state   (b) Left-right symmetry   (c) P1-P2 symmetry

Fig. 6. Symmetric states in Chinese Checkers.

Definition 4. *A game of Chinese Checkers is drawn if any board state is repeated during play.*

Under this rule, if a drawn position is reached players can either repeat the position to draw the game, or a player can move into a position that is a theoretical loss, hoping that the opponent will make a mistake later in game play. While other games require positions to be repeated multiple times to cause a draw, it seems that one repetition should be sufficient.

3.5 Chinese Checkers Symmetry

Chinese Checkers has two types of symmetry. The first is right-left symmetry where flipping the board results in a position that must have an equivalent value to the original position. This type of symmetry is shown in Fig. 6(a) and 6(b). This symmetry was exploited in previous work when solving the single-agent version of Chinese Checkers [19], but also applies to the two-player version of the game. While the symmetry can be used to reduce the size of the stored solution to the game, the symmetric states will still be reached during the process of solving the game. Note also that there are some positions which are identical when flipped, and thus the savings from this symmetry is close to, but does not reach a factor of 2. On the 6×6 board with 6 pieces our implementation reduces the number of stored states by a factor of 1.998.

Chinese Checkers also has symmetry between player 1 and player 2. That is, if we flip the board top-to-bottom and swap the colors and the player to move, we also end up in a symmetric position. The state in Fig. 6(a) is symmetric to the state in Fig. 6(c).

4 Strongly Solving the Chinese Checkers

We have built a number of different solvers that strongly solve the game of Chinese Checkers in different ways, including in-memory solvers and external-memory solvers. The various solvers have been used to test the efficiency of solving techniques and to verify the correctness of each of the new implementations.

Many of the structural choices have been based on our earlier work studying the influence of solving parameters [17]. One focus of the work has been strongly solving the game efficiently using moderate hardware, as opposed to using massive parallelization to solve the game more quickly, yet less efficiently (from a CPU/power usage perspective).

Three important choices in the solving process are (1) the ranking function which orders the states in the state space (2) the order in which states are proven, and (3) immediate propagation of wins to parent states once a win is proven. On 7 × 7 Chinese Checkers with 3 pieces per player and well-optimized choices for these parameters, we can determine the value of 99% of the states in the state space in one pass through the data, and only 8 passes through the data are needed to complete the proof. Without parent propagation or an optimized proof ordering, the proof takes up to 34 passes through the data with only 0.5% of the states proven in the first iteration. Both proofs find the same result, but the first implementation performs the computation more efficiently.

5   Results

Results from our solves are found in Table 1. We have solved a variety of board sizes of Chinese Checkers with up to the 6 × 6 board with 6 pieces per player, which has 2 trillion states. The solvers use two bits per state, so given the two types of symmetry, this solution requires storing 500 billion states and 135 GB of storage, which exceeds the main memory available in the computer used for building the solution. As a result, external memory (disk) was used for this solution. A full description of this solver is outside of the scope of this paper, but our solver borrows ideas from external memory BFS implementations such as search with structured duplicate detection [20] and TBBFS [8].

Table 1. Proven results and win/loss/draw/illegal counts from various size games. All results except for the 6 × 6 board with 6 pieces have been solved multiple times with different solvers to validate the results.

Board Size	# Pieces	Positions	Wins	Draws	Illegal	Start
7 × 7	1	4,704	2,304	0	96	P1 Win
7 × 7	2	2,542,512	1,265,851	0	10,810	P1 Win
7 × 7	3	559,352,640	279,297,470	180,860	576,840	P1 Win
7 × 7	4	63,136,929,240	31,532,340,944	51,686,042	20,561,310	P1 Win
4 × 4	6	3,363,360	1,205,441	547,058	405,420	P1 Win
5 × 5	6	9,610,154,400	4,749,618,788	47,056,118	63,860,706	P1 Win
6 × 6	6	2,313,100,389,600	1,153,000,938,173	5,199,820,604	1,898,692,650	P1 Win
7 × 7	6	170,503,381,976,928	?	?	?	?

Because the game is symmetric between the two players, the number of wins and losses is the same. Thus, in Table 1 we only report the number of won states in the game. These counts are for the full game, ignoring symmetry. Note that all results, except for the largest game, were computed multiple times and by multiple solvers, validating the counts. We are in the process of building a more efficient external-memory solver which will validate these results with significantly better performance. The exact count of wins should not be considered correct until this verification has been completed.

Our goal is to strongly solve the 7×7 board with 6 pieces per player, as this is the largest game that can likely be strongly solved on hardware that we have readily accessible.

6 Conclusions and Future Work

This paper has outlined the steps required to begin strongly solving games of Chinese Checkers, including rules for ending the game and for drawn states. The games solved thus far are all first-player wins, something we do not expect to change on larger board sizes. Besides providing new information about a game that has not previously been solved, the solved information provides a new testbed for studying games.

References

1. Allis, L.V.: Searching for Solutions in Games and Artificial Intelligence. Ph.D. thesis, Maastricht University (1994)
2. Allis, V.: A knowledge-based approach of Connect-Four-the game is solved: White wins. Ph.D. thesis, Vrije Universiteit (1988)
3. Björnsson, Y., Schaeffer, J., Sturtevant, N.R.: Partial information endgame databases. In: van den Herik, H.J., Hsu, S.-C., Hsu, T.-S., Donkers, H.H.L.M.J. (eds.) ACG 2005. LNCS, vol. 4250, pp. 11–22. Springer, Heidelberg (2006). https://doi.org/10.1007/11922155_2
4. Carlisle, R.P.: Encyclopedia of Play in Today's Society, vol. 1. Sage, Thousand Oaks (2009)
5. Edelkamp, S., Kissmann, P.: Symbolic classification of general two-player games. In: Dengel, A.R., Berns, K., Breuel, T.M., Bomarius, F., Roth-Berghofer, T.R. (eds.) KI 2008. LNCS (LNAI), vol. 5243, pp. 185–192. Springer, Heidelberg (2008). https://doi.org/10.1007/978-3-540-85845-4_23
6. Henderson, P., Arneson, B., Hayward, R.B.: Solving 8x8 hex. In: Twenty-First International Joint Conference on Artificial Intelligence (2009)
7. Irving, G.: https://perfect-pentago.net/details.html, https://perfect-pentago.net/details.html
8. Korf, R.E.: Minimizing disk i/o in two-bit breadth-first search. In: AAAI Conference on Artificial Intelligence, pp. 317–324 (2008)
9. Romein, J.W., Bal, H.E.: Solving awari with parallel retrograde analysis. Computer **36**(10), 26–33 (2003)
10. Roschke, M., Sturtevant, N.R.: UCT enhancements in chinese checkers using an endgame database. In: Cazenave, T., Winands, M.H.M., Iida, H. (eds.) CGW 2013. CCIS, vol. 408, pp. 57–70. Springer, Cham (2014). https://doi.org/10.1007/978-3-319-05428-5_5

11. Schadd, M.P., Winands, M.H.: Best reply search for multiplayer games. IEEE Trans. Comput. Intell. AI Games **3**(1), 57–66 (2011)
12. Schaeffer, J., Björnsson, Y., Burch, N., Lake, R., Lu, P., Sutphen, S.: Building the checkers 10-piece endgame databases. In: Van Den Herik, H.J., Iida, H., Heinz, E.A. (eds.) Advances in Computer Games. ITIFIP, vol. 135, pp. 193–210. Springer, Boston, MA (2004). https://doi.org/10.1007/978-0-387-35706-5_13
13. Schaeffer, J., et al.: Checkers is solved. Science **317**(5844), 1518–1522 (2007)
14. Silver, D., et al.: A general reinforcement learning algorithm that masters chess, shogi, and go through self-play. Science **362**(6419), 1140–1144 (2018)
15. Sturtevant, N.: Challenges and progress on using large lossy endgame databases in chinese checkers. In: IJCAI Workshop on Computer Games (2015)
16. Sturtevant, N.: A comparison of algorithms for multi-player games. In: Schaeffer, J., Müller, M., Björnsson, Y. (eds.) CG 2002. LNCS, vol. 2883, pp. 108–122. Springer, Heidelberg (2003). https://doi.org/10.1007/978-3-540-40031-8_8
17. Sturtevant, N.R., Saffidine, A.: A study of forward versus backwards endgame solvers with results in chinese checkers. In: Cazenave, T., Winands, M.H.M., Saffidine, A. (eds.) CGW 2017. CCIS, vol. 818, pp. 121–136. Springer, Cham (2018). https://doi.org/10.1007/978-3-319-75931-9_9
18. Sturtevant, N.: Multi-player games: Algorithms and approaches. Ph.D. thesis, UCLA (2003). http://www.cs.du.edu/~sturtevant/papers/multiplayergames thesis.pdf
19. Sturtevant, N., Rutherford, M.: Minimizing writes in parallel external memory search. In: International Joint Conference on Artificial Intelligence (IJCAI), pp. 666–673 (2013). http://www.cs.du.edu/~sturtevant/papers/bfs_min_write.pdf
20. Zhou, R., Hansen, E.A.: Parallel structured duplicate detection. In: Twenty-Second AAAI Conference on Artificial Intelligence (AAAI), pp. 1217–1224. AAAI Press, Vancouver, British Columbia, Canada, Jul 2007. http://www.aaai.org/Library/AAAI/2007/aaai07-193.php

A Practical Introduction to the Ludii General Game System

Cameron Browne$^{(\boxtimes)}$, Matthew Stephenson, Éric Piette,
and Dennis J. N. J. Soemers

Department of Data Science and Knowledge Engineering, Maastricht University,
Bouillonstraat 8-10, 6211 LH Maastricht, The Netherlands
{cameron.browne,matthew.stephenson,eric.piette,
dennis.soemers}@maastrichtuniversity.nl

Abstract. Ludii is a new general game system, currently under development, which aims to support a wider range of games than existing systems and approaches. It is being developed primarily for the task of game design, but offers a number of other potential benefits for game and AI researchers, professionals and hobbyists. This paper is based on an interactive demonstration of Ludii at thuis year's Advances in Computer Games conference (ACG 2019). It describes the approach behind Ludii, how it works, how it is used, and what it can potentially do.

Keywords: General game system · General game playing · Game description language · Ludeme · Ludii · Game design · Artificial intelligence

1 Introduction

Ludii is a *general game system* (GGS) [4] for modelling, playing, evaluating, optimising, reconstructing and generating a range of games in a digital format. It is distinct from existing GGSs in that its primary purpose is as a game design tool, with the focus being on the flexibility and expressiveness of its design language and the ease with which games can be defined.

1.1 The Digital Ludeme Project

Ludii is being developed as part of the Digital Ludeme Project (DLP)[1], a five-year research project which aims to model the world's traditional strategy games in a single, playable digital database. This database will be used to find relationships between games and their components, in order to develop a model for the evolution of games throughout recorded human history and to chart their spread across cultures worldwide. This project will establish a new field of research called Digital Archæoludology [2].

---

[1] Digital Ludeme Project: http://ludeme.eu/.

© Springer Nature Switzerland AG 2020
T. Cazenave et al. (Eds.): ACG 2019, LNCS 12516, pp. 167–179, 2020.
https://doi.org/10.1007/978-3-030-65883-0_14

The DLP will model the thousand most influential traditional strategy games throughout history, each of which may have multiple interpretations and require hundreds of variant rule sets to be tested. These will mostly be board games but will also include card games, dice games, tile games, etc., and will involve games with non-deterministic elements of chance or hidden information. The Ludii system was developed for this purpose, as no existing general game approach would support the full range of games required for the execution of the DLP.

The following sections describe the approach behind Ludii, its game grammar and compilation mechanisms, how games are represented and played, how the user interacts with the system, and potential services Ludii might offer.

2   Ludemic Approach

Ludii is based on a *ludemic approach* that decomposes games into atomic constituents describing relevant equipment and rules.

2.1   Ludemes

Ludemes are "game memes" or units of game-related information that represent the building blocks of games; they are the conceptual units that game designers work with when developing their designs. The term was coined in the 1970s by Alain Borvo for his analysis of a novel card game [1].

The following example shows how the game of Tic-Tac-Toe might be described in ludemic form. All information required to play the game – the players, the equipment and the rules – are presented in a simple, structured format:

```
(game "Tic-Tac-Toe"
   (players 2)
   (equipment {
      (board (square 3))
      (piece "Nought" P1)
      (piece "Cross" P2)
   })
   (rules
      (play (to (empty)))
      (end (if (line 3) (result Mover Win)))
   )
)
```

The ludemic approach is a high-level approach to game description that encapsulates the key game-related concepts while hiding the complexity of the underlying implementation, making it well suited to the task of game description and design. This is in contrast with existing approaches, such as the Stanford Game Description Language (GDL) [13], that explicitly state the instructions for updating the game state in the descriptions themselves, yielding verbose and complex descriptions that do not encapsulate relevant concepts and are less amenable to the modifications required for game design.

2.2 Ludi

Ludii is based on similar principles to the first author's previous Ludi game system, which was used to evolve combinatorial board games in ludemic form [5]. However, Ludii has been completely redesigned to address shortcomings in its previous incarnation, in order to provide the generality, extensibility and efficiency required for the successful execution of the DLP. These improvements are due mainly to the class grammar approach for automated game grammar generation, and Monte Carlo-based move planning with a forward model only, yielding speed-ups in the order of 100 times faster for most games.

3 Class Grammar

The Ludi *class grammar* is a set of production rules derived directly from the Java code implementation of the ludeme classes, in which sequences of symbols on the RHS are assigned to a nonterminal symbol on the LHS very much like an Extended Backus-Naur Form (EBNF) grammar [6]. The basic syntax is as follows:

```
<class> ::= { (class [{<arg>}]) | <subClass> | terminal }
```

where:

`<class>`	denotes a LHS symbol that maps to a class in the code library.
`(class [{<arg>}])`	denotes a `class` constructor and its arguments.
`Terminal`	denotes a terminal symbol (fundamental data type or `enum`).
`{...}`	denotes a collection of one or more items.
`|`	denotes a choice between options in the RHS sequence.

The grammar is intrinsically bound to the underlying code library, but is *context-free* in that it is self-contained and can be used without knowledge of the underlying code. The mechanism for generating the grammar is similar to that of parsing C++ constructors described by Hall [11] to produce a form of *domain specific language* (DSL) [10]. The Ludii class grammar is effectively a snapshot of the class hierarchy of the program's current ludeme code base in Java. Ludii is implemented in Java for its cross-platform support, performance, flexible compilation and good Reflection library support.

3.1 Annotations

Custom annotations are used to decorate arguments in ludeme class constructors to help shape the resulting grammar. For example, the `@Opt` annotation is used to denote optional arguments for a ludeme, @Named is used to denote arguments that must be named in the grammar, and @Or denotes consecutive runs of arguments of which exactly one must be specified in the game description. For example, a Line class constructor with the following signature:

```
public Line(
            @Name final IntFunction length,
        @Opt       final Direction   dirn,
    @Or @Opt @Name final IntFunction what,
    @Or @Opt       final RoleType    who
)
```

would generate the following rule with named and optional arguments:

```
<line> ::= (line length:<int> [<direction>] [(what:<int> |
<roleType>)])
```

3.2 Game Descriptions

Games are described as *symbolic expressions* or *s-expressions* expressed in the Ludii class grammar. The following example shows the game of Havannah, in which players win by connecting two corners, three board sides (not including corners) or form a ring with their pieces:

```
(game "Havannah"
    (players 2)
    (equipment {(board (hexagon 8) (hexagonal)) (piece "Ball"
Each)})
    (rules
        (play (to (empty)))
        (end
            (if (or {(connect 2 Corners)(connect 3
SidesNoCorners)(ring)})
            (result Mover Win)
        )
    )
)
```

3.3 Game Compilation

Game descriptions are processed using a *recursive descent parser* [7] in which LHS class names are matched to the actual classes they refer to. The (terminal or non-terminal) arguments to each (non-terminal) class are compiled, then the appropriate class constructor is found, compiled and passed up the compilation hierarchy. The object returned at the root of this compilation process is an executable Game object ready to run.

3.4 Advantages and Disadvantages

Advantages of the class grammar approach include its easy extensibility, as any required functionality can be simply implemented, added to the code base, and automatically subsumed into the grammar. The system will theoretically support any functionality that can be implemented in Java, taking a step towards the ideal of the programming language becoming the game description language [15].

A drawback is that users adding ludemes to the code base must follow strict formatting guidelines for the ludeme constructors if they are to produce a well-behaved grammar. However, these are well documented for those who need them.

4 Game Representation

A game in Ludii is given by a 4-tuple $= \langle Players, Mode, Equipment, Rules \rangle$. $Players$ is a finite set of k players described by the numbers of players. $Mode$ is the type of the game between: Alternating (by default if not specified), Simultaneous and Real Time. $Equipment$ describes the $containers$ and the $components$ of the game. The containers are mainly a description of the main board by its shape and its tiling and if necessary the hands of the players. Each component is described by the ludeme $piece$ specifying its name, its owner and if necessary how this component can be moved in the board. Finally, $Rules$ defines the operations of the game which is split in three distinct parts: $start$, $play$ and end.

For each container, the system builds a graph representation of the board according to its tiling and precomputes any useful data structure (neighbours of each vertex, corners of the board, etc.) in order to efficiently compute the legal moves from each game state.

4.1 Game States

When a game is compiled different flags corresponding to game types are automatically generated in function of the ludemic description. According to them, a game state in Ludii is built. A set of ContainerState objects associated with each container defines a game state. A Container state is defined using a custom BitSet class (called ChunkSet) that compresses the required state information into a minimal memory footprint. The ChunkSet encodes multiple data: What(locn) the index of the component located at locn, Who(locn) the owner of this piece, count(locn) the number of this component, an internal state of a component (direction, side, etc.) by state(locn) and if the information hidden to a player are given by hidden(player, locn).

4.2 Moves and Actions

Legal moves are described by a Moves object which contains a list of component Move objects generated by the "play" rules of the game for the given state. Each move equates to a *complex instruction set* (CISC) command that decomposes into a set of atomic *reduced instruction set* (RISC) `Action` objects, each of which typically modifies a `ChunkSet` in the state.

For example, the move `To(1, 4)` sets the piece with index 1 at cell location 4 of the default container (i.e. the board), by applying the sequence of atomic actions: { `SetWhat(4,1)`, `SetWho(4,1)` }.

5   AI Agents

One of the primary aims of Ludii is to facilitate the implementation of general game playing agents, and their evaluation in a wide variety of game. To this end, Ludii contains a number of default agent implementations, and provides an interface for the development of third-party agents.

5.1   Default AI Agents

The default agents implemented in Ludii are:

- **Random**: an agent that samples actions uniformly at random.
- **Monte-Carlo (flat)**: an agent that estimates the values of actions available in the root node using a flat Monte-Carlo search (i.e. uniformly random playouts), and selects the action with the maximum estimated value.
- **UCT**: a standard UCT implementation [3,9,12]. An open-loop Monte-Carlo tree search (MCTS) approach [14] is used in stochastic games.
- **MC-GRAVE**: an implementation of Generalized Rapid Action Value Estimation [8].
- **Biased MCTS**: a variant of MCTS that uses simple patterns as features for state-action pairs to bias [16] the selection and playout phases of MCTS.

5.2   Third-Party AI Support

Ludii provides an interface for the implementation of new agents, which can subsequently be imported into Ludii's GUI and used to play any Ludii game. Programmatic access to Ludii's game is also available, which allows for convenient evaluation of custom algorithms using Ludii's wide array of implemented games. Example implementations are available on github.[2]

6   Ludii Player

In this section we describe the Ludii player, that provides the front-end interface for accessing the complete functionality of the Ludii system. Some of the main highlights of the Ludii Player include:

---

[2] https://github.com/Ludeme/LudiiExampleAI.

- A graphical interface for playing hundreds of traditional and modern strategy games, both locally and online internationally, with other players from around the world.
- A variety of included general game playing algorithms (UCT, Flat-MC GRAVE, etc.) with comprehensive evaluation metrics and visualisation options, as well as the ability to integrate third-party agents.
- Tools for creating, playtesting, and sharing your own game designs, defined using the Ludii general game language.

6.1 Game Playing

One of the key advantages of Ludii over other previous general game playing systems such as GGP, is the ability to view and interact with all implemented games via a sophisticated graphical environment. This allows human users to play and enjoy any game created within Ludii, whilst also making tasks such as correcting bugs and identifying incorrect rule descriptions much easier to perform. The main graphical interface provided by the Ludii player is shown in Fig. 1. The left side of the view shows the main playing area of the current game, the top right section provides information about the games's players (name, colour, score, components in hand, etc.), and the bottom right area provides additional information about the game (moves made, game description, AI analysis, etc.).

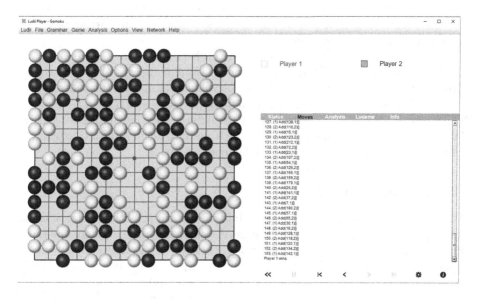

Fig. 1. The Ludii Player interface, showing a completed game of Gomoku.

The Ludii player currently allows up to 8 players (both human and AI) to play games against each other, either on a single Ludii application or else using multiple applications within a single local network. By registering for a free Ludii user account, Ludii games can also be played internationally with other human players online. A large and active community of players from many different demographics and geographic regions, may also provide valuable insight into the game playing preferences and abilities of different cultures.

The Ludii Player also includes many customisable options for improving the overall game experience. Examples include changing the colours of the board and pieces, visualising the game's mathematical graph, showing the axes or coordinates of the game board, displaying the possible moves that can be made by each player, providing a complete list of all moves made, the ability to undo moves, and analysis on the current winning likelihood of each player.

6.2   Agent Evaluation

As well as allowing humans to play games, the Ludii Player also contains several features that make it easier to evaluate and analyse the performance of different agents. Any player within a game can be controlled by one of the provided game playing algorithms that are included with Ludii. Analysis provided by these agents, such as their iteration count, child mode visits, and value estimates, is provided directly within the Ludii Player. It is also possible to visualise the distribution of possible moves for each agent at any given game state, see Fig. 2, providing a graphical representation of the AI"thought process". Moves that involve adding a new piece into the game are represented by a dot, see Fig. 2a, whilst those that involve changing the position of an existing piece are represented by arrows, see Fig. 2b.

The size of either the dot or arrow for each possible move represents how much the AI is thinking about that move (playout number), whilst the colour indicates the average score obtained from playouts after making this move (red = low win likelihood, blue = high win likelihood, purple = neutral win likelihood). These features makes it easier to identify the different playing abilities of general game playing agents, and can also provide a useful teaching tool for explaining how certain algorithms search the available action space.

6.3   Manual Game Creation

The Ludii player also provides many useful tools for aiding with the creation and testing of games using the Ludii language. A complete game editor for the Ludii Player is currently under development and will allow designers to adjust certain properties or ludemes of the current game's description directly within the Ludii app, with the resulting changes being compiled and applied automatically.

A large number of board and piece designs will be included within the Ludii player for game designers to use – see Fig. 3 for examples – but it will also be possible to specify your own piece and board images within Ludii game descriptions. This will provide a wide range of possibilities for both the rules and visuals of a created game. These game descriptions can the be loaded into any Ludii

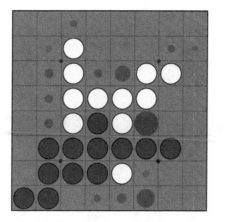

(a) Reversi - black to move.   (b) Chess - white to move.

Fig. 2. AI visualisations for two example games (Reversi and Chess) showing the outcome likelihood (colour) and number of playouts (size) for each move.

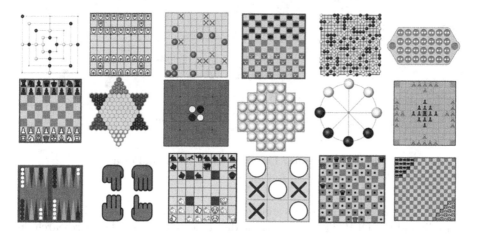

Fig. 3. Thumbnails for some of the games provided with the Ludii Player.

application, allowing designers to easily share their created games with other Ludii users.

7 Ludii Portal

The Ludii Portal website, hosted at the URL www.ludii.games, provides additional information and services beyond those offered by the main Ludii system. Some of these services are not yet available at the time of writing, but will be added to the Ludii Portal over the coming months.

Library. The Ludii Game Library provides a wide range of computational and historical information on the complete collection of official Ludii games, see Fig. 4. This includes diagrams, rule descriptions, strategies, tutorials, mathematical and social profiles, geographical regions, time periods, cultural importance, game reconstructions, and much more.

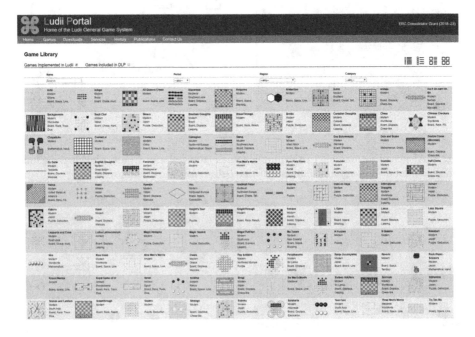

Fig. 4. The Game Library page of the Ludii Portal.

Forum. The Ludii Forum offers a dedicated space to discuss any subject related to Ludii and the DLP. This may include discussions about the latest research on general game AI, archaeological finds from recent excavations, promotion and sharing of new game descriptions, the results of Ludii competitions, recommendations for new games to include within the official Ludii game repository, and whatever else the Ludii community feels is important to discuss.

Competitions. We plan to run several general game AI competitions using Ludii over the following years [17]. This includes many AI competitions that focus on the development of autonomous general game playing agents, procedural content generators, and data mining algorithms. In addition to this, we aim to organise a handful of non-AI related competitions focusing on human playing abilities and game design. Such competitions will hopefully stimulate conversation on the Ludii forum, and would likely rely on the cooperation of a large number of Ludii users to compete, playtest and evaluate submitted entries.

Game Recommendations. Another service that will be offered by the Ludii Portal will be personalised game recommendations. These recommendations will be based on user personal information, prior game results within the Ludii Player, regional and cultural data, and other factors that may influence an individual's game preference. The more users that participate in this game recommendation system, the more accurate our suggestions will be.

8   Planned Services

Ludii provides a platform for many potential game design services, including the following, which we plan to provide over the course of the DLP.

8.1   Automated Game Design

Games might be generated automatically in a number of modes:

- **Unconstrained:** New games might be generated through standard search techniques (e.g.. evolutionary methods or hill-climbing techniques) using the provided database of known games as a starting point.
- **Directed:** New games might be generated by directed search according to metrics, conditions or desired behaviours specified by the user.
- **Bespoke:** Games might be generated for individual users based on implicit preferences inferred from player behaviour.

8.2   Game Optimisation

Ludii has already proven to be a useful tool for automated play-testing to detect imbalances and other flaws in candidate rule sets. This may in future be coupled with intelligent rule modification to optimse rule sets in order to reduce or (ideally) remove flaws.

8.3   Historical Game Reconstruction

One of the most important services offered by Ludii will be the facility to perform reconstructions of historical games based on partial or unreliable information. This includes taking material evidence in the form of (possibly partial) game boards and pieces, and inferring likely rule sets based on the geographical, historical and cultural context of the evidence, using historical data accumulated during the course of the DLP through archival and on-site research [2]. The aim is to produce likely reconstructions that maximise historical authenticity as well as quality of play, and to provide a tool to help traditional games researchers in the difficult reconstruction process.

9   Conclusion

The Ludii general game system, while being developed to address the needs of the larger Digital Ludeme Project, has the potential to be a significant and useful software tool in its own right. It has been designed to allow the description of as wide a range of (mostly traditional) games as easily as possible, and to provide a platform for a range of game analysis and design services that games researchers will hopefully benefit from. Ludii will continue to mature and expand in functionality as the DLP progresses.

Acknowledgements. This research is part of the European Research Council-funded Digital Ludeme Project (ERC Consolidator Grant #771292) run by Cameron Browne at Maastricht University's Department of Data Science and Knowledge Engineering.

References

1. Borvo, A.: Anatomie D'un Jeu de Cartes: L'Aluette ou le Jeu de Vache. Librarie Nantaise Yves Vachon, Nantes (1977)
2. Browne, C.: AI for ancient games. KI - Künstliche Intelligenz **34**(1), 89–93 (2019). https://doi.org/10.1007/s13218-019-00600-6
3. Browne, C., et al.: A survey of monte carlo tree search methods. IEEE Trans. Comput. Intell. AI Games **4**(1), 1–49 (2012)
4. Browne, C., Togelius, J., Sturtevant, N.: Guest editorial: general games. IEEE Trans. Comput. Intell. AI Games **6**(4), 1–3 (2014)
5. Browne, C.B.: Automatic Generation and Evaluation of Recombination Games. Phd thesis, Faculty of Information Technology, Queensland University of Technology, Queensland, Australia (2009)
6. Browne, C.: A class grammar for general games. In: Plaat, A., Kosters, W., van den Herik, J. (eds.) CG 2016. LNCS, vol. 10068, pp. 167–182. Springer, Cham (2016). https://doi.org/10.1007/978-3-319-50935-8_16
7. Burge, W.H.: Recursive Programming Techniques. Addison-Wesley, Boston (1975)
8. Cazenave, T.: Generalized rapid action value estimation. In: Yang, Q., Woolridge, M. (eds.) Proceedings of the Twenty-Fourth International Joint Conference on Artificial Intelligence (IJCAI 2015), pp. 754–760. AAAI Press (2015)
9. Coulom, R.: Efficient selectivity and backup operators in Monte-Carlo tree search. In: van den Herik, H.J., Ciancarini, P., Donkers, H.H.L.M. (eds.) Computers and Games. LNCS, vol. 4630, pp. 72–83. Springer, Berlin Heidelberg (2007)
10. Fowler, M., Parsons, R.: Domain-Specific Languages. Addison-Wesley, Boston (2011)
11. Hall, P.W.: Parsing with C++ constructors. ACM SIGPLAN Not. **28**(4), 67–69 (1993)
12. Kocsis, L., Szepesvári, C.: Bandit based monte-carlo planning. In: Fürnkranz, J., Scheffer, T., Spiliopoulou, M. (eds.) ECML 2006. LNCS (LNAI), vol. 4212, pp. 282–293. Springer, Heidelberg (2006). https://doi.org/10.1007/11871842_29
13. Love, N., Hinrichs, T., Genesereth, M.: General game playing: Game description language specification. Technical report LG-2006-01, Stanford Logic Group (2008)
14. Perez, D., Dieskau, J., Hünermund, M., Mostaghim, S., Lucas, S.M.: Open loop search for general video game playing. In: Proceedings of the Genetic and Evolutionary Computation Conference, pp. 337–344. ACM (2015)

15. Schaul, T., Togelius, J., Schmidhuber, J.: Measuring intelligence through games. CoRR abs/1109.1314 (2011). http://arxiv.org/abs/1109.1314
16. Soemers, D.J.N.J., Piette, É., Browne, C.: Biasing MCTS with features for general games. In: Proceedings of the 2019 IEEE Congress on Evolutionary Computation (CEC 2019), pp. 442–449 (2019)
17. Stephenson, M., Piette, É., Soemers, D.J.N.J., Browne, C.: Ludii as a competition platform. In: Proceedings of the 2019 IEEE Conference on Games (COG 2019), pp. 634–641. London (2019)

Author Index

United States